FRM

WHERE

I SIT

*Essays on Bees,
Beekeeping, and Science*

Mark L. Winston
Department of Biological Sciences
Simon Fraser University

Comstock Publishing Associates,
A DIVISION OF
Cornell University Press
ITHACA AND LONDON

First published 1998 by Cornell University Press.
First printing, Cornell Paperbacks, 1998.

Printed in the United States of America.

Library of Congress Cataloging-in-Publication Data

Winston, Mark L.
    From where I sit : essays on bees, beekeeping, and science / Mark
L. Winston.
        p.   cm.
    ISBN 0-8014-3477-7 (cloth : alk. paper). — ISBN 0-8014-8478-2
(pbk. : alk. paper)
    1. Honeybee. 2. Bee culture. I. Title.
    SF523.3.W55 1998
    638'.1—dc21   97-41016

Cloth printing   10 9 8 7 6 5 4 3 2 1

Paperback printing   10 9 8 7 6 5 4 3 2 1

Contents

Foreword

BEES PROVIDE A RICH SOURCE OF INTEREST TO everyone concerned with nature and science. Beekeepers themselves are endlessly fascinated by bees, the honey they produce, and their pollinating activity, which has a significant impact on the North American agricultural economy. In these essays Mark Winston uses bees to bridge the gap between scientists and the public, and to demonstrate how scientists work—and the importance to everyone of scientific research. At the same time he strongly encourages scientists to become more accountable to the society that pays their salaries. These entertaining essays will inform and stimulate many readers besides beekeepers—naturalists, gardeners, farmers, researchers in other subjects—to think more deeply about bees, science, and nature.

A senior researcher with wide experience in the United States and Canada, Mark Winston has a very alert mind which he now turns to important problems facing beekeepers and bee researchers in these countries. He explores issues not often discussed by the beekeeping community. No subject that affects beekeeping, the honey industry, or bee research is too controversial for him to bring into the open. *From Where I Sit* ranges over the USDA, education, research funding and publication, legislation, university financing and faculty appointments, and relationships between organizations—and between individuals. In each essay Professor Winston calls on his own experience, and when he is critical of an existing system, he makes suggestions for improving it.

Winston regards the aims of publicly funded bee research as improving the beekeeping industry and increasing the financial return from it. He believes projects for bee research paid for by a government or university should be examined for their practical use to beekeepers, and that beekeepers should be involved in funding decisions.

Mark Winston's essays will be of special interest in North America—the focus of his concern—and will give beekeepers and researchers on other continents a lively overview of the character of present beekeeping and bee research in the United States and Canada. Taken together, these essays provide a valuable snapshot of the present stage of apiculture in these two technologically advanced countries.

Winston also goes beyond the details of research to express wonder at many features of the honey bee, especially its social life. In this, he follows a line of distinguished writers from Aristotle and Virgil in the ancient world to Montaigne and Shakespeare in the 1500s. The honey bee referred to in these essays is the European *Apis mellifera*, native to the Mediterranean region where early civilizations flourished. (There are at least seven other *Apis* species, all less amenable to management.) For more than four thousand years this European bee has been fostered by man and kept in hives. Since 1600 beekeepers have spread it to almost all places in the world where plants grow. The bee prospered so well that it became the source of most of the world's honey and the pollinator of many commercial crops. At the end of the essay "Life in the Research Lane," Winston quotes the words inscribed on a border arch between the United States and Canada: "Children of a common mother." These same words could well be applied to all the descendants of the original European honey bee that today live in different parts of the world. In many regions these bees now face danger on several fronts. Winston examines the threats and describes the measures being taken to combat them.

The author uses beekeeping and bee research as a basis for exploring much broader issues. If he stimulates readers to think more about the function of apiculture within the increasingly complex social and environmental constraints beekeepers now face, he will have succeeded in his objective.

—EVA CRANE
formerly Director, International Bee Research Association

Acknowledgments

I AM PARTICULARLY GRATEFUL TO KIM FLOTTUM, the editor of Bee Culture magazine, who first suggested I write a monthly column. I had no idea concerning the direction or tone these columns would take when I began writing in 1993, and I continue to be surprised at what emerges each month. Kim's confidence in my writing, and willingness to take chances by allowing me to write about issues not usually discussed in the beekeeping community, has opened up an unexpected aspect to my career. I continue to look forward to writing my column each month, and hope to keep writing the columns as long as the excitement remains.

I also appreciate the careful editing and keen eye for detail that my editor at Cornell University Press, Peter Prescott, brought to this book. Most of these articles are similar to the original versions that appeared in *Bee Culture*, but Peter's advice as to which articles to choose and how to organize them, and his subtle improvements in the writing, have made this book accessible to a wider readership. The manuscript also benefited from detailed and perceptive editing by Mindy Conner and Helene Maddux.

In addition, it is a pleasure to acknowledge my own mentors, colleagues, and especially students. I have been fortunate in receiving good guidance and caring examples to follow in my career from those who taught me the crafts of science and beekeeping. I also have had the pleasure of working with many of the scientists who are actively pushing the limits of what we know and how we communicate it. Most treasured, however, have been my own students, those who took my courses and the high school, undergraduate, and graduate students and postdoctoral fellows who have worked in my laboratory. No experience has meant more to me than being able to teach others, and nothing has taught me more than pondering the questions and observing the growth of students.

Finally, I thank my friends, colleagues, and correspondents in the beekeeping community around the world. I have been privileged to travel extensively to speak about bees, and the warm generosity and keen interest of beekeepers everywhere has been a source of real pleasure. I am particularly grateful to the beekeepers of British Columbia and Canada who have opened their apiaries, shared their ideas, and welcomed me, my family, and my students into their homes. I also value beyond words my colleagues in the Canadian Association of Professional Apiculturists, whose collegiality, advice, and friendship mean more to me than they can ever know.

M. L. W.

From Where I Sit

Introduction

Essay: A short literary composition on a single subject, usually presenting the personal view of the author.
—*Nelson Canadian Dictionary*, 1997

THE ESSAYS IN THIS BOOK WERE SELECTED FROM the monthly columns I write for the beekeeping magazine *Bee Culture*. They were written between 1993 and 1997, and represent a snapshot of what was happening in bee biology, management, politics, and lore during that time. They go further, however, in presenting my own perspective on bees and, through bees, on issues as far-ranging as nature, science, teaching, and especially the role of scientists in society.

These were not issues that had always intrigued me. Like most academic scientists, my early career was focused on obtaining grants and tenure, writing scientific papers, trying to keep up with my teaching load, and making time for a life outside of work. I was fortunate in choosing to study bees, because as I surfaced from the strain of building a career I discovered that bees and bee research could be much more than simply a focus for my job. In many ways, bees are a paradigm for understanding the role science should play in society and for exploring the myriad ways humans interact with, understand, and manage nature.

It is not difficult to comprehend why someone would choose to study bees. They are fascinating organisms. The purely social nature of the honey bee, their amazing dance communication system, the reign of the queen and the subservice of the workers, the adaptations of colonies to environments as diverse as sweltering tropical jungles and frigid temperate forests: all these and much more have been the subjects of innumerable studies, papers, books, movies, and intense speculation.

Beyond interesting is the economic importance of bees, and it is on the path between curiosity and management that most of this book travels. We have exploited bees for honey and wax since the dawn of human existence, and in the last 150 years have developed sophisticated management systems to ensure that bees produce copious quantities of honey and pollinate billions of dollars worth of crops every year. There are thought to be over 6 million beekeepers worldwide, and the number of managed colonies easily reaches 50–100 million around the globe. The diverse ways we keep bees are a testimony to human ingenuity and to the role science can play in improving the human condition, because it is through scientific research that we have developed the management techniques that allow us to exploit the utility of this versatile organism.

The 1990s have been among the most compelling of times to study bees in North America. Beekeepers have been faced with many challenges, and scientific research has played a major role in meeting these challenges. The issues facing beekeepers—and the solutions developed by scientists—have been reported extensively in the media. Such problems as the Africanized "killer" bees, which threaten public health and bee management, and the arrival of tracheal and *Varroa* mites in North America have been among the prominent issues that have interested the public and engaged bee researchers in looking for solutions.

These issues have stimulated scientists to pay more attention to the imperative, contemporary need for applied research, and have forced many apiculturists to consider how to balance our role as basic researchers with the demands for relevance coming from the beekeeping community. For me, my *Bee Culture* column has provided an opportunity to consider the broader issues behind the research subjects being pursued today. Putting this book together has made me aware that my monthly musings have fallen into themes, and that through my column I have developed a perspective on the role researchers can and should play in society.

If a single theme emerges from this book, it is this: The most enriching aspect of research is not new discoveries, although investigations arising from curiosity will always remain at the core of science. Rather, it is at the intersection of discovery and utility, the point where basic science and applied science connect, that our endeavors can attain their most glorious triumphs. Our most serious obligation as scientists is to return to society some of the value invested in our work, and to move beyond the selfish joy of discovering new things to fulfill our promise as contributors to the greater good.

This may appear self-evident, and even to me reads like a platitude, except for one thing: too many of my colleagues consider applications of their research as an afterthought, and working with the private sector an occupational hazard rather than one of the real joys of the job. My experience with researchers is that too often we move on when the basic studies are done, and fail to perform the flip side: applied research that provides tangible justification for society to fund our curiosity.

Beyond the narrow confines of bee research, I continue to be surprised at how unwillingly some scientists communicate to the public, and at how token our justifications for research funding can be. Many of us lack interest in the civic arena and exhibit poorly developed interpersonal skills when we are forced to interact with the public. Not enough of us relish the interactions with industry, agriculture, and the media that bring our work before the public eye. In this sense, bees and bee research exemplify a pervasive problem with academic and government research communities, a problem I have tried to highlight and address in many of these essays.

Another theme that emerged in choosing the essays to include in this collection is how bee science mirrors broader issues facing society today. Some of these essays explore topics such as why we prefer pesticides for pest management, and why alternative, environmentally benign treatments have not been able to replace harsh chemical methods. Other essays consider the enormous impact imported pests have had in North America, and the reasons these importations have been and continue to be so problematic. Another theme is the importance of giving the public a role in making decisions about what science to fund, and giving real decision-making power to public groups in focusing research dollars on the needs of society.

I have also attempted to profile how scientists work, and to address the diverse issues that captivate contemporary scientists. Science remains a fascinating life-style, regardless of how well we scientists implement and communicate our results, and the complexity of modern science has added new layers of complexity to our studies. Modern molecular techniques enable us to probe the genome, and we have begun to learn how gene expression is translated into behavioral action. Our knowledge about behavior has also expanded dramatically in recent years, and our appreciation of social organisms like the honey bee has increased concomitantly. Philosophically, bees are a probe with which to explore the human condition and our connection with nature, and in some of these essays I have tried to show how contemporary bee research has changed the way we perceive the natural world and our place in it.

Underlying all these themes are bees themselves, a rich lode of adaptations that teach us about nature and reflect back to us some of what it is to be human. Perhaps this book could have been written about almost any species, but the social nature of the honey bee makes it a particularly compelling exemplar of broader issues. Bees have led me to consider topics as diverse as religion, funding, pesticides, behavior, and disease, all stimulated by the different perspectives bees offer to those who take the time to think about them.

Ultimately, writing these essays has given me the opportunity to engage socially relevant issues while contemplating the beauty and complexity of nature. I am fortunate in being able to pursue a career in which I can enter the world of another creature, the honey bee, and immerse myself in the magnificent adaptations and stunning behaviors of this well-studied but still largely unknown organism. Beyond wonder has been utility, and it is the balance between marveling at nature and describing research useful for my own species that has made writing about the honey bee and beekeeping such a delight.

PART ONE
Thinking about Bees

Thinking about Bees

Honey bees are social, they sting, and they produce honey. That alone should be enough to stimulate us to think about them. We marvel at the high level of social interaction that holds the colony together as a unit of tightly coordinated individuals, a model from which our human societies might benefit. We fear the sting of the worker bee with an innate repulsion because it is painful and for some of us can be fatal. We covet their honey, and the earliest art painted on cave walls depicts our ancestors plundering wild hives for their sweet contents.

These are all good reasons to study and speculate about the honey bee, but it is also self-interest that leads us to explore the lives of these distant animal relatives. We force them to live in wooden hives and manipulate them to collect copious quantities of nectar from flowers and pollinate many of our crops. Yet, even these bees in hives are essentially still wild; they can inhabit hollow tree trunks in forests as easily as manufactured hives on farms.

Beekeeping is an extensive human occupation. It occurs on all continents except Antarctica, and millions of people keep bees as a hobby or profession. Honey bees are unusual in this sense of being both wild and domesticated, in existing in more or less the same form in natural situations and managed hives. We have considerable contact with other domesticated animals—cows, pigs, sheep, chickens, cats, and dogs—but none of these retain the wild attributes still found in honey bees; nor do they survive well outside the domesticated state. Honey bees are different. They bridge the gap between human and nature, between feral and managed, and they remind us of our own natural roots.

Even city dwellers are interested in bees. There is an extensive subculture of urban beekeepers who keep bees in backyards, on garage roofs, and nestled on apartment balconies. Apprehensive neighbors are placated with a bucket of honey, and regaled with tales of the hive from their beekeeping friends. Bees pollinate city gardens, fly on their own airborne roads back and forth from their collecting jobs, and bring a sense of the rural to the urban environment. They inspire urbanites to think about bees and nature, and to remember their own often distant rural ancestry.

Similarly, people have an insatiable appetite for bee lore, and for watching bees and learning about their individual and collective functions in the hive. Is there a major city in North America that does not have a science museum or park with a glass-walled observation hive stocked with bees as its most popular exhibit? Here in Vancouver, British Columbia, my students manage the observation hive in Science World. Invariably the bees are surrounded by a throng of children and adults searching for the queen, watching the workers perform their tasks, and releasing a barrage of questions about what they are seeing, or not seeing. I have often seen individuals transfixed by the hive, an almost meditative state of involvement with the colony. Surely this fascination with bees has value.

People also have a special interest in the hind end of the bee, where the sting resides. Unfortunately, bees get an undeserved bad rap from the public in this regard. Most "bee" stings are more correctly attributed to wasps, which are considerably more aggressive and more likely to sting than bees. Almost everyone has been stung by a wasp or bee at some point in their lives, and almost everyone remembers it. I cannot count the number of times, on buses or planes, at parties or at work, that I have been asked what I do for a living and then had to listen to a long, painfully detailed, and invariably boring story about a sting. Yet, the cumulative weight of years of these stories has made me respect the impact a sting can have on its recipient.

The public's panic at the spread of Africanized "killer" bees is a further example of the intensity of humans' aversion to stinging organisms. In 1956, South African honey bees were imported to Brazil in an attempt to breed a more manageable bee for tropical beekeeping. Some of the African bees escaped into the wild and have spread throughout the tropical and subtropical regions of the New World. They are more aggressive than the European-derived honey bees we are accustomed to in North America, and our media have been full of gruesome tales from Latin America about mass stinging attacks by the Africanized bees.

The arrival of Africanized bees in the southwestern United States in 1991 elicited a new round of killer bee stories, movies, and tabloid journalism. I've seen movies in which these bees destroy a nuclear missile facility and the city of Houston, invade the domed stadium in New Orleans, and imprison and terrorize a family in their California home. The continued production of these poorly made, badly acted, and terribly inaccurate movies can only be explained by a subliminal human fascination with bees and their sting, a fear deeply imbedded in our genome as a survival tool that balances our interest in bees with a healthy respect for the danger they represent.

We also are becoming increasingly aware of other bees in addition to the honey bee. There are many thousands of bee species, each with its own adaptations for survival. Some live as solitary individuals, and others exhibit social behavior as complex as the honey bee's. These bees have existed largely unnoticed in our fields and forests, but our interest in them has increased as large-scale agriculture and urban development have diminished their populations. The arrival of tracheal and *Varroa* mites and the resulting decline in managed and feral honey bee colonies has been the immediate stimulus for this attention to other bee species. Consequently, scientists are searching for alternative bee species to use in pollination management, and we are becoming more cognizant of and interested in the other wild bee species.

There is another level of interest in bees that usually is not discussed around the beekeeping world, or anywhere else for that matter: the religious and metaphysical. Bees appear repeatedly in religions as symbols and icons, and art depicting bees is an unrecognized but active form of expression. Bees mean something to people, something beyond their stings, beyond the memories of rural farms they evoke or the delight we feel at seeing nature flying among skyscrapers and between suburban homes. There is a depth to bees, a sense of something greater than ourselves; a connection to the divine lies at the heart of our fascination with these insects. If we open and explore a colony of honey bees and tune out the noise of traffic, forget scientific debates, ignore what we know about their behavior, and just feel the presence of this sublime organism, something happens. It is very much like prayer.

1. Bees in the City

I ALWAYS KNOW WHAT SEASON IT IS, WITHOUT HAVing to look outside, simply by the type of telephone calls I get from the residents of Vancouver, British Columbia, where I live and work. The barrage of calls about bees and wasps begins on the first warm spring day with the predictable "I found a killer bee in my house, what should I do?" Late spring is characterized by "There's a bee nest on my front porch, and I heard you study bees and was wondering if you would like to take it away and study it?" I know August has arrived when the calls change to "We have a wasp nest on our house, and I heard you study wasps and was wondering if you would like to take it away and study it?" Early September is marked by calls from the media, each year with the same emphatic question: "This summer was the worst for wasps in anyone's memory. Why were there so many more wasps this year?"

The spring of 1996, however, was marked by a different type of call. For the first time, the callers were asking "What's happened to all the bees? I haven't seen any honey bees in my garden or on my fruit trees and berry bushes. Will they get pollinated this year?" I dismissed the first call or two as nonsense, since the spring had been our finest in many years, but when I kept getting such calls day after day, I finally realized it was a trend, and perhaps there really were fewer honey bees in the city.

Such a shortage had never happened before, because urban beekeeping has long been a popular hobby in Vancouver, as it is in most

North American cities. When I moved to my Vancouver home, the eighty-year-old gentleman across the street maintained a colony of bees on his flat garage roof, in full view of the street. His neighbor had ten colonies in his backyard, stocked with swarms he located by leaving his name with the local police and fire departments, who passed on calls about swarms in the neighborhood. A fellow who lived on the busy main street three streets over maintained three colonies of bees on his second-floor balcony, facing the street. My two backyard colonies barely swelled the neighborhood bee population.

Urban beekeeping is not only common, it can also be very successful. My own backyard colonies produce at least one hundred pounds of honey each year, more than the bee yards out in the country average. One Vancouver beekeeper used to brag about the twenty colonies he maintained on a rooftop in the downtown area, and about how each colony produced four-hundred-pound harvests annually. These harvests gave him guru-like status in his local bee club until everyone realized that his bees were located within a few hundred yards of a sugar-packing facility!

Most cities regulate beekeeping with a hodgepodge of bylaws. Vancouver, for instance, prohibits beekeeping within the city limits, although public health officials don't enforce the law unless someone complains. Indeed, there is a very public and obvious Vancouver Bee Club, and beekeepers maintain hives at the municipally run public garden plots. Nearby municipalities allow beekeeping in a properly fenced backyard, as long as the bees don't cause a nuisance, but hiding bees isn't really necessary, no matter what the regulations may be. I can't imagine that the police haven't noticed the bees from the three colonies on the front balcony a few streets over from us. No, beekeeping generally is tolerated within the city, and any objections by the neighbors seem to disappear after receiving a bucket or two of honey.

Managed honey bees in the city provide a major public service by pollinating gardens, fruit trees, and berry bushes, and should be encouraged rather than legislated out of existence. Our cities, groomed and cosmopolitan as they appear, still obey the basic rules of nature, and our gardens and yards are no exception. Home-grown squashes, apple trees, raspberries, peas, beans, and many other garden crops require bees to move the pollen from one flower to another, no matter how urbanized or sophisticated the neighborhood. The demise of wild bees due to pesticide use and destruction of nest habitat has made backyard gardens even more dependent on managed pollinators than many agricultural areas, and any problem with urban beekeeping resonates through our backyards.

The apparent decrease in Vancouver honey bees that my callers reported was probably due to parasitic tracheal and *Varroa* mites. Beekeepers in my region were hit hard when *Varroa* mites arrived in 1993 and added their effects to those of the already present tracheal mites. The Ministry of Agriculture reported a 66 percent loss in bee colonies the first season, and there was a tremendous shortage of bees for pollination in the agricultural regions of the nearby Fraser Valley. The impact of these mites has been most severe on hobby beekeepers, many of whom are not involved in organized bee clubs, don't get bee journals and newsletters, and are largely unaware of the management techniques necessary to prevent mite damage to colonies. And even those who do know don't always have the resources, time, or interest to perform the intensive treatments necessary to manage bees in today's mite-infested world. I think the drop in honey bee numbers reported by my callers is a real one, and it does not bode well for gardens and plantings in urban environments.

Urban beekeepers in the southern United States are beginning to confront another "pest" that will curtail beekeeping in the city, the Africanized bee. These bees are not compatible with urban beekeeping and should be kept a few hundred yards from people, pets, homes, schools, and roads. While city beekeeping is still possible in the post-Africanized world by maintaining colonies with European queens, I expect to see a decline in urban beekeeping as the Africanized bees spread. Indeed, fear of legal action following a stinging incident may lead to stricter enforcement of municipal bylaws, and outright bans on bees in some cases and urban hobby beekeepers' lack of knowledge about how to keep colonies from becoming Africanized will conspire to reduce managed bee populations in our cities. Even in Canada, where Africanized bees are unlikely to arrive, municipalities have been tightening their rules about beekeeping in response to news reports about Africanized bees from the southern United States.

Another threat to urban beekeeping is pollution. Bees are notorious for concentrating pollutants in their honey, largely because many pollutants wind up in floral nectar before the bees collect it. Plants that breathe in airborne pollutants or take up waterborne substances through their roots tend to concentrate these compounds in their flowers, leading to high levels of pollutants in nectar. Airborne substances may be trapped in bee hives by the waxy comb and sticky honey, and are easily carried by bees themselves on their hairs. Bees near highways or foraging on flowers along heavily traveled roadsides may be especially likely to concentrate pollutants.

Pesticides are another problem for city bees, just as they are for country bees. A few years ago we discovered large numbers of twitching, dying bees at the entrance to one of our backyard colonies. I had the bees analyzed by the Ministry of Agriculture, and they tested positive for Diazinon, a pesticide used frequently by backyard gardeners. Interestingly, only one of our two colonies was affected, likely because its bees were foraging in an area where blooming plants had been sprayed while those from the other colony were not. There is little that a beekeeper can do to protect colonies from backyard pesticides because there may be hundreds or thousands of homes within foraging distance of an urban bee colony.

A final threat to urban beekeeping is angry neighbors, but it is rarely the threat of stings that upsets them. No, the most fervent complaints about backyard bees are due to orange fecal droppings that fall on freshly washed cars or newly painted porches. I also know spring has arrived by the annual round of calls from city officials attempting to mediate between an irate citizen complaining about bee feces and the beekeeper. Most of these problems can be easily solved by relocating the bees elsewhere on the property, or sometimes just by offering to wash the neighbor's car every week or two during the spring.

When I finally get away from my phone and get the chance to go outside with my bees, I rediscover the greatest joy of backyard beekeeping: my neighbors. Rather than feeling threatened by my bees, my neighbors are intensely curious, and we have had many conversations over the fence that began with bee talk and ended with our becoming much better acquainted. Neighbors do occasionally fall out over bees, but more often bees bring people together.

Backyard beekeeping is a great conversation starter, and also a re-minder that no matter how urbanized we have become, nature still flows and flies in the city. It will be a great loss for all urban dwellers, not just beekeepers, if mites, Africanized bees, pesticides, and pollution drive bees out of the cities. There is a bit of farmer and country dweller in even the most urbane city resident, and bees in the city are an important timekeeper of the seasons that flow on away from the concrete, skyscrapers, and freeways that dominate urban landscapes.

2. Feral Bees

I RECENTLY BEGAN DOING SOME READING ON A SUB-ject I call "nature thought" to provide some background for a book I'm writing. The literature is vast, and extraordinarily dense and convoluted, but it examines such important issues as whether humans are part of nature or separate from it, and our relationships with the rest of the earth's inhabitants. Discussions on humans and nature go back to the Old Testament and beyond, and the issues haven't changed much in the last five thousands years. Basically, humans have spent much of recorded history trying to decide whether we are stewards or dominators of nature. As stewards, we have attempted to foster the natural world and maintain it, but as dominators we have used plants, animals, and entire ecosystems for our own benefit. There is little doubt that the dominator philosophy is winning, in spite of an increasingly strong and active pro-environment movement. Pastoral ideas of stewardship are simply overwhelmed by the food and resource imperatives of six billion voracious humans.

Nevertheless, the resilience of the natural world astounds me, and this is especially true of the partly domesticated honey bee. Compare the honey bee with any other domesticated species, such as the cow. There is no such thing as a feral cow, and cows have been so highly selected and domesticated for our needs that the ancestral cow would be almost unrecognizable. Indeed, our cattle would die if we were no longer here to provide food and shelter for them, treat their diseases, and protect them from predators. In contrast, honey bees don't depend on humans for survival. Our disappearance from earth might have the opposite effect on them, with feral honey bee colonies increasing and thriving in our absence.

Feral and domesticated honey bees do differ in subtle ways, however, for each has been subjected to different selective forces that have molded their divergent characteristics. The recent Africanized bee invasion into the New World has forced us to focus attention on unmanaged, natural honey bee colonies, and we now know quite a bit about feral honey bees and how they both differ from and interact with managed colonies.

Managed bees were derived from wild colonies. The natural variation that exists in wild colonies provided the substrate for selection and enabled humans to produce a much more usable and productive bee. Many centuries of domestication and interbreeding between populations from distant geographic areas, including Europe, have created a North American managed bee whose characteristics differ from those of feral tree-living populations. The degree of separation between feral and managed bees is narrow, however, because feral and managed queens and drones mate and interchange characteristics, swarms from managed colonies escape into the feral population, and hobby beekeepers sometimes stock new hives with captured feral swarms or harvested wild colonies. The two populations exist in an uneasy balance. Beekeepers select for large colonies with docile, relatively nonswarming traits, while natural selection pushes for feral colonies that are smaller, swarm more frequently, and are more aggressive.

Recent studies using the techniques of molecular biology have confirmed that the genetic makeup of feral and managed bees differs. Nathan Schiff and Walter Sheppard in particular have shown that the commercial bee population in the United States is strongly dominated by Carniolan and Italian bees (*Apis mellifera carnica* and *A. m. ligustica*), while the feral population has a relatively high proportion of the German dark bee (*Apis mellifera*), which was imported through the nineteenth century but is no longer considered desirable for beekeeping. Further, feral bees show considerable variation throughout the United States, which is not surprising given the wide range of climatic and habitat conditions encountered by feral colonies, while managed bees tend to be more homogeneous in their genetic makeup.

Commercial beekeepers traditionally have considered feral colonies a nuisance at best, because feral bees are a reservoir for many diseases, compete for forage with managed colonies, and interfere with attempts to control queen mating. Even hobbyists are now discouraged from obtaining bees from feral sources, partly because they carry diseases, but also because the characteristics of feral bees do not make for easily managed colonies. A number of events occurring in the beekeeping world today, however, are forcing us to pay more attention to feral colonies.

This point was brought home to me in a reply by Richard Taylor to a letter published in the August 1994 issue of *Bee Culture*. The letter inquired about feral bees, and Taylor's reply suggested that the feral population has been greatly reduced by the parasitic *Varroa* and tracheal mites. He went on to say that the feral colonies that remain tend to be mite resistant and could thus serve as a gene reservoir useful for beekeeping. Some beekeepers have tried a managed version of this large natural experiment by not treating colonies in their apiaries with miticides and then breeding from queens in the surviving colonies. The real-world version is more selective, however, because feral colonies do not have the survival advantages that supplemental feeding, well-constructed hives, protection from winter, transportation to good forage, and antibiotic treatments give to managed colonies.

Mite resistance in feral bees could also be detrimental for commercial beekeeping, however. There were probably millions of feral colonies in the southern United States before tracheal and *Varroa* mites arrived, and the niche created by colony deaths due to mites could be filled by mite-resistant bees with other, undesirable characteristics. The name of one potential feral replacement is a familiar one: the Africanized bee. There is some evidence that *Varroa* mites are not as damaging to Africanized bees as to European-derived races, possibly because the shorter development time of Africanized workers does not give the *Varroa* mites sufficient time to reproduce before the adult bee emerges. Optimistic predictions of a minimal impact of Africanized bees in the United States require northward-spreading Africanized bees to encounter high populations of feral and managed European bees in the southern and western United States. But the death of many if not most feral colonies because of mites, coupled with reduced numbers of managed colonies due to mite damage and economic factors, could prevent the predicted amelioration and provide an opportunity for Africanized bees to become well established.

Thus, the future feral bee population poses interesting dilemmas for the steward and dominator sides in the human-nature interaction. On the one hand, feral bees might provide important genetic stock for use in future breeding programs to minimize mite or other pest damage in managed colonies. On the other hand, the feral population may be highly undesirable for other reasons such as carrying Africanized traits. The beneficial aspects of mite resistance might be overwhelmed by the negative impact of aggressive bees that swarm often and don't grow to large populations in colonies even when managed. The stewards among us would want to maintain feral bees, while the dominators would opt to eliminate feral bees, if possible, and leave the natural and agricultural worlds to be served by the managed population.

Fortunately, we probably won't have to face this dilemma, because we haven't been able to influence feral bees to any significant extent.

Our failure to have any impact on the spread or characteristics of Africanized bees is good evidence that the beekeeping community has not yet domesticated honey bees to the point that dairy and cattle farmers have domesticated cows. Bees are still basically feral organisms that respond more strongly to natural selection than to human selection. We have domesticated honey bees to the extent that we can manage them, but we have not yet destroyed their ability to survive in nature.

I'm comforted by this because I think there is some merit to a world in which humans don't always win over nature. In many ways I appreciate the fact that the Africanized bee has succeeded in overcoming all our attempts to modify its characteristics and stop its spread through the tropical Americas, and it also pleases me that feral bees in North America are different from managed ones. Anyone who has cut open a bee tree or caught a wild swarm knows the sense of wonder and admiration that accompany the realization that something as complicated as a bee colony can survive without human help. I'm certainly not naive enough to think that modern agriculture could prosper without managed bees, but I'm enough of a steward to appreciate the feral population out there that does not respond to our needs or evolve for our benefit. Sure, let's do our best to select managed bees with characteristics that are good for beekeeping, but let us also appreciate the feral bees that don't, and shouldn't, answer our call.

3. Feral Bees II

IN THE PRECEDING ESSAY I WROTE ABOUT FERAL honey bees, but there is another type of feral bee that is becoming the object of increasingly nervous attention from the beekeeping community. I'm referring, of course, to the lowly bumble bee and its relatives. Wild bees have cohabited in nature with honey bees for many millions of years, and in North America for a few hundred years, ever since honey bees were imported from the Old World. Until recently, their existence and health have not been of much interest or concern to any but the most dedicated bee aficionados. Wild bees have always been interesting to bee scientists and collectors, but today they are a hot topic because they have the potential to replace honey bees as managed pollinators in some commercial pollination situations.

It is surprising that it took us so long to realize that bees other than honey bees can serve as commercial pollinators. There are more than twenty thousand species of bees worldwide, many of them well adapted to pollinate particular crops because they evolved to visit a narrow spectrum of food plants. This specialization makes them better pollinators of certain plants than honey bees, which visit a broad range of plants but don't do a particularly good job of pollinating many of them. Many wild bee species are socially inclined, or at least nest in aggregations, and can thus be grouped together to provide the large number of individual bees necessary to perform commercial pollination.

Three wild bee species have been successfully domesticated as managed pollinators in addition to the honey bee. The alfalfa leaf-cutting bee has been the most successful field crop pollinator, and is used routinely in the alfalfa seed production industry. These bees are better pollinators of alfalfa than honey bees because alfalfa flowers "trip" when visited by a bee, and this action seems to prevent honey bees from effectively transferring their pollen. The leafcutting bee has no difficulty learning to cope with a tripping flower, and so does a more effective job of pollinating. These bees are easily managed, and can be kept indoors and cool during the winter, and warmed up and set out in the millions the next spring when the alfalfa is blooming. The liberated adult females return to the nesting straws set out for them, provision cells, and lay eggs, so the beekeeper can repeat the cycle.

The orchard bee is another managed pollinator, although its success has been limited to Japan. Orchard bees also make cells in straws, which can be collected and stored over the winter and then set out in apple orchards the following spring. Commercial apple pollination in Japan is done almost exclusively by these bees rather than by honey bees, but orchard bees have not been adopted in North America, even though intensive research here suggests they would be commercially viable. The management systems orchard bees require may be too labor-intensive for large-scale mechanized North American agriculture. Also, there seem to be problems with restricting the orchard bees to the apple crop and getting them to return to nest in the boxes provided for them. Whatever the reasons, these bees remain a potential but as yet unrealized managed pollinator in North America.

Bumble bees, the most recently managed wild bee pollinators, have become the major pollinator for some greenhouse crops, especially tomatoes. Methods for rearing bumble bees were developed by Chris Plowright and Cam Jay in Manitoba during the 1960s, but were not applied commercially until the mid 1980s, when several Dutch companies began producing bumble bees in large quantities. The techniques for mass rearing bumble bees are now well known, although the methodology is a closely guarded trade secret and is not readily available to beekeepers. Initially, the rental fee for a single bumble bee colony was as much as six hundred dollars per colony—and even at that was viewed as reasonable in an industry where pollination had previously been accomplished by hand. The price dropped dramatically with increases in colony production, however, and the current price is less than two hundred dollars per colony. Numerous studies have shown that bumble bees are highly cost-effective under glass, and do a superb job of pollinating some greenhouse crops. This bumble bee industry has taken off, and a number of companies have amassed considerable profits through selling bumble bee colonies to greenhouse growers.

I think it is unfortunate that beekeepers are not the ones making money from these alternative pollinators; we, after all, have the skills and background to rear bees. But beekeepers seem curiously reluctant to support research into alternative pollinators, viewing these "other bees" as competitors for their pollinating income rather than as an opportunity to diversify and make more income. I think the beekeeping community needs to develop a different perspective about alternative pollinators. Beekeepers need to realize that wild bees will never replace the honey bee as the most important broad-spectrum managed pollinator, and their income from pollination is not in great danger of being supplanted by these alternative bees. Nevertheless, these bees are significant supplemental or even replacement pollinators for a few crops, and beekeepers should be the ones at the forefront of research and management. These opportunities for alternative pollinators should be exploited by astute beekeepers rather than by astute businesspeople.

It is important to remember that the honey bee will continue to dominate the pollination industry. Wild bees simply do not have the potential to take over the important role honey bees play in the contemporary agricultural community. Honey bee colonies have the overwhelming advantage that they can be nurtured to large populations, and these populous hives can be moved to crops during bloom. Although feral bees might provide tens or hundreds of bees per acre, and do a better job pollinating a crop's flowers as individuals, a single honey bee colony can provide thousands or even tens of thousands of foraging workers, each of which does at least an adequate job of pollinating. Further, honey bee colonies can easily be transported thousands of miles from one flowering crop to another, so the same colony can be used three or even four times a season. Thus, beekeepers should remain confident in continued income from honey bee pollination, and use that confidence as the basis to explore new pollinating situations.

Domesticated wild bees have the potential to provide enhanced pollination services in a few markets. I'll stick my neck out here and make a prediction: bumble bees and leafcutting bees will make up about 10 to 20 percent of the berry pollination business in the northeast and northwest United States, and the equivalent regions in southern Canada, within ten years. These regions have experienced a shortage of honey bee colonies in recent years because of tracheal and *Varroa* mites, and the rules of supply and demand have come into play. There are too few colonies to pollinate crops such as blueberry and cranberry, and the price paid per colony has risen dramatically, so expensive alternative pollinators have become a viable alternative. My region, for example, experienced a 66 percent colony loss in 1993, and berry growers imported honey bee colonies from distant regions and paid twice the price per colony they had paid the previous year. In this economic environment, bumble bee colonies have become almost equivalent to honey bees in cost-effectiveness, and the new suppliers certain to arise in this situation will drive the price of bumble bee colonies down even further. Some berry growers are now purchasing bumble bee colonies as insurance against a continued shortage of honey bees, and the use of bumble bees will continue to increase. Although bumble bees will not replace honey bees in this system, there seems to be good potential for them to take on an increasingly important role in commercial berry pollination. Experiments also have begun with leafcutting bees on cranberry, with some success.

My model for tomorrow's commercial pollinating beekeeper is a fellow I met in New Zealand a few years ago who made most of his income from renting his thousand or so honey bee colonies to kiwifruit growers. Kiwifruit is not very attractive to honey bees, so alternative pollinators have some potential in this market. He proudly showed me a small incubating room in his warehouse that he was filling with bumble bee nests. He had been experimenting with bumble bees for many years and was finally ready to begin renting colonies to some very interested growers. I looked at the stacks of honey bee boxes that went from floor to ceiling in rows stretching the length of his warehouse, and then looked at the small walk-in incubator where he kept his bumble bees, and did some quick calculations. The potential income from the bumble bees in this small incubator was almost the same as the income from all the honey bee colonies represented by the hives stacked in his warehouse. And further, the physical labor involved in raising bumble bees can easily be done by a child, while managing a thousand-hive honey bee operation took his full attention and hard labor and that of his sons throughout the year. His diversification may result in additional income by providing a more effective multispecies pollinating service to the kiwifruit growers, but, most important, the income from this service will be staying in the beekeeping community.

Rather than fearing the onset of alternative pollinators or taking the extreme view that they will replace honey bees, beekeepers should look on other pollinators as opportunities for diversification and supplemental income. If we don't, be assured that someone else will, and I'm sure that you, like me, would rather see the beekeepers making money, no matter what bee species they are managing.

4. Death, Where Is Thy Sting?

A LETTER I RECEIVED FROM A BROOKLYN, NEW York, couple recently reminded me that some people truly dislike being stung. The couple was criticizing an article I wrote a few years ago for the *Encyclopedia Britannica* about Africanized bees. I attempted to take a balanced approach in this article, emphasizing that the importance of bees for honey production and pollination made it imperative that North Americans learn to manage and live with Africanized bees. I concluded by saying, "People must learn to adjust to a new, successful, and highly adaptable organism in their midst." To that my correspondents responded: "Professor Winston, in spite of all his knowledge about bees, must have sustained one too many bee stings himself, resulting in some brain damage." They went on to say that "if some of the so-called good honey bees also have to be exterminated along with them, so be it and good riddance to the killer bees. Believe me, the human race will somehow learn to survive with less pollination of crops."

My first reaction to this letter was to dismiss it as coming from the city-dwelling fringe element, but that same week I received a paper from the *Quarterly Review of Medicine*, an English academic journal, that described the symptoms, reactions, and attempted treatment of five males who had been the victims of mass stinging attacks by Africanized bees in Brazil. The clinical language and excruciating medical detail of the article were frightening even to me, and suddenly I began to see the Africanized bee from the perspective of the nonbeekeeping public. Beekeepers brush off stings as part of our livelihood, but for most people, stings elicit the same terror as shark attacks, snake bites, and Stephen King movies.

Honey bee stings are a marvelous adaptation for defense, at least from the bees' point of view. Honey bees are unusual among insects in that the stinger remains imbedded in the victim and the bee dies. The stinger is composed of two barbed lancets supported by hard plates and powerful muscles and connected to a venom sac and to specialized glands that produce alarm odors. When a bee stings, the lancets scissor their way into the victim, and the barbs anchor the sting so that it remains in the skin when the bee pulls away. The sting continues to throb for thirty to sixty seconds, injecting venom and giving off alarm odors that alert other bees and mark the victim for continued attack. Presumably, the additional venom that is injected when the sting is left in the victim, and the alarm odors that remain behind to attract additional attackers compensate for the loss of the bee's life. In colonies with many thousands of workers, the loss of a few during nest defense is thus balanced by having a more potent and effective sting.

The venom is a blend of compounds that are highly effective against a wide range of potential attackers. The major component is a protein called melittin, but venom also contains other compounds such as histamine. The complex nature of the venom may be an adaptation to the wide variety of insect and vertebrate pests and predators that can attack bee colonies; different components of the venom seem to be important in repelling different species of attackers. For example, the amount of histamine in a bee sting is not sufficient to be toxic to vertebrates but is toxic against other insects, including other honey bees. Each of the major venom components has somewhat different effects on vertebrates; their sum we see as an allergic reaction.

Humans react to stings on three levels: local, systemic, and anaphylactic. In a local reaction, the initial localized swelling is followed by more extensive swelling a few hours later, and the affected area may be red, itchy, and tender for two to three days. A systemic reaction generally occurs within a few minutes of stinging, and may involve a whole-body rash, wheezing, nausea, vomiting, abdominal pain, and fainting. Symptoms of an anaphylactic reaction can occur within seconds and include difficulty in breathing, confusion, vomiting, and falling blood pressure, which can lead to loss of consciousness and death from circulatory and respiratory collapse.

Single stings by Africanized bees and the European-derived bees now used in North America are almost identical in their effects. It is at the colony level that the two differ: Africanized colonies are likely to respond to minimal disturbances with sudden large-scale attacks. Although some colonies of European bees sting readily and some Africanized colonies are gentle, the average Africanized colony is considerably more aggressive than most European colonies. Furthermore, the extreme attacks that can occur with Africanized bees are almost unknown in European bees. An attack by Africanized bees can quickly escalate into a mass stinging incident in which one or more victims may be frequently stung and pursued for more than a kilometer by attacking bees. The victim can receive hundreds or thousands of stings within minutes, leading to death from a large-scale systemic reaction.

It was such large-scale systemic reactions that were the subject of the *Quarterly Journal of Medicine* article. Although unusual, whole-body reactions are truly scary. They involve an almost total breakdown of organ systems leading to death. Typically, the victims were admitted to São Paulo hospitals after receiving two hundred to more than one thousand stings. The article used such frightening medical terms as *intravascular haemolysis, respiratory distress, hepatic dysfunction, hypertension, myocardial damage, shock, coma*, and *acute renal failure* to describe the victims' symptoms. In regular language, the victims received toxic doses of venom that caused the heart, lungs, blood, and urinary systems to fail within twenty-four to forty-eight hours.

These reactions were caused not by allergies but by high doses of venom from the large number of stings received. Even more frightening, three of five victims studied died in spite of receiving sophisticated treatment with antihistamines, corticosteroids, bronchodilators, vasodilators, intravenous saline, bicarbonate, mannitol, and mechanical ventilation. In other words, prompt, state-of-the-art medical attention was not sufficient to save them.

How common are mass stinging incidents? It is difficult to determine the precise number of human deaths caused by Africanized bees, but the few statistics available indicate an increase in fatality rates of five to ten times or more over the pre-Africanized levels. About 400 people died from excessive stinging in Venezuela between 1975 and 1990, and about 240 deaths have been reported in Mexico since 1989. By 1996, there had been 3 fatalities in Texas following mass stinging from Africanized bees. Generally, fatality rates have been highest during the first few years of Africanization, and most of those who died were elderly individuals. Some stinging incidents can be spectacular; one individual in Costa Rica was stung more than eight thousand times, an average of seven stings per square centimeter.

Unfortunately, these stinging incidents, unusual though they are, have the potential to seriously damage beekeeping in the United States because they play to people's worst fears about bees. As responsible beekeepers, we need to take these events seriously and do what we can to both minimize the possibility of mass stinging and demystify bees so that humans' innate fear of bees can be overcome by familiarity with this beneficial insect.

Reducing the likelihood of mass stinging is the easiest aspect of Africanized bee public relations to accomplish. Good beekeeping is consistent with reduced stinging. For example, annual requeening with non-Africanized queens will eliminate most of the problems Africanized bees can cause in managed colonies, and choosing isolated apiary sites and rapidly eliminating Africanized colonies from beeyards will reduce contacts between people and Africanized bees.

I think we need to do more, however, to promote the image of bees with the general public. Contact and familiarity are the best ways to reduce people's fears and encourage an image of bees not as "killers," but rather as life-givers through their roles in pollination and honey production. To that end, I like some of the wackier things beekeepers do to get publicity, such as bee beards and queen-finding contests. Although these activities may seem frivolous, they do bring people closer to bees, and that proximity leads to familiarity and loss of fear. We often perform mass bee beardings at Western Apicultural Society conferences. Although most of the onlookers stand far away when we start, at the end there are invariably hundreds of spectators mingling with the beard wearers, no longer afraid and truly interested in these marvelous insects. Two of our University secretaries put on bee beards at a recent conference at Simon Fraser University, and they proudly show off their beard photographs to everyone who visits the conference center.

Another superb way to get the message across is in the schools. Observation hives are very effective for reducing fear and encouraging interest in bees. Pennsylvania State University developed a program that took school visits one step further; they invited teachers to the university campus for a two-day program that demonstrated how honey bees can be used for classroom teaching. Is there a better science lesson than training bees to forage at a hidden dish containing sugar syrup and then having the students interpret the dance language and find the dish? At the end of the program, each teacher was given an observation hive stocked with bees to take back to his or her school.

Nevertheless, serious stinging incidents will happen in spite of all our efforts at good public relations, and we should also encourage the development of an effective antivenom. The sting kits available today are effective against allergic reactions but do not provide an adequate response to the large quantities of toxins injected in a mass stinging incident. An antivenom specific for bee stings would have saved the lives of the Brazilian victims who died because the available medical technology was not effective.

Aggressive public education, attention to safety with our bees, and the development of an effective antivenom are all excellent ways to diminish the real and perceived impact of bees on the public. Africanized bees can create serious difficulties for beekeepers, but perhaps the most difficult dilemma the bees cause is a public relations problem. Now that the bees are in the United States, we need to do more to promote bees as interesting and important insects. Stop for a minute the next time you're routinely working your bees, marvel at the extraordinary adaptations and behaviors that make them one of nature's miracles, then take that feeling and head right over to the nearest school and start sharing that wonder with the kids. It will be the best thing you can do to maintain the viability of our industry.

5. Bee Brains

Each of us has many identities, different aspects of our personalities that are expressed in different situations. We may be one person in church or synagogue, another at work, another tossing down a beer after work, and still another person at home. My colleagues at work may consider me an intellectual interested in the pursuit of knowledge and study, someone who behaves in a very "professional" way. They would be surprised to get in my car with me after work as I turn on one of my many country music tapes and listen to Tanya Tucker whine about heartbreak or sing along with Travis Tritt about how the whisky ain't workin' anymore.

I began thinking about this while flying back from Paris one August after attending a meeting of social insect biologists at which one of the major topics was bee brains. The meeting was held at the Sorbonne, one of the oldest universities in the world. In hallowed halls surrounded by intellectual history, the walls covered with oil paintings depicting the great thinkers of the past, we listened to lectures delivered in theaters where centuries ago French scientists presented their work without the aid of slide projectors and pointers, with only their brains and their ability to think and speak as tools.

This meeting allowed me to express one of my personalities that beekeepers might find odd, especially those who regularly read my *Bee Culture* column and have figured out that I think scientists should do at least some practical research, and should be able to communicate what they do simply and clearly. I listened to talk after talk about basic bee biology that had virtually no practical relevance to beekeepers, and very little potential of ever leading to commercially useful applications. Even worse, I heard some of the most respected scientists in the world present talks so technical and so convoluted that it took all my concentration to even begin to understand them. Nevertheless, I enjoyed myself because I kept thinking about the world from a bee's point of view.

I learned enough at this meeting to know that calling someone a "bee brain" is a major compliment rather than a mild insult. The brain of a worker bee is about the size of a large pinhead, yet bees are able to use their brains to behave differently in different contexts, much as my academic persona shifts to my country music personality when I leave work. I already knew that bees could learn, that they have memories, and that individuals make decisions about which job to do depending on the colony's requirements. At these meetings I learned that we are beginning to understand not only what bees can do, but how their brains and hormonal systems control simple mechanisms that can lead to complex behavioral decisions.

One of the most complicated things a worker bee has to do is make its way to flowers, determine how to extract the nectar or pollen, and then find its way home. And this after the bee has spent its life until then inside the nest doing tasks such as brood rearing or comb building that impart little or no information about how to forage. It's rather like raising a girl inside her home without ever allowing her outside the yard, then suddenly, when she becomes a teenager, sending her to the supermarket a few miles away to find sugar on the shelf, figure out how to buy it, and then make her way home.

A naive forager requires all sorts of new information to survive outside, find food, and successfully return to the nest. The young worker bee's brain is fine for directing the tasks the bee does inside the nest but is not physically adequate to accommodate these new skills. It must undergo physical changes that enable it to accommodate the information needed to forage. A study conducted here in North America by Susan Fahrbach, Gene Robinson, and their student Ginger Withers at the University of Illinois has shown that the bee's brain grows before the bee begins to forage, probably to provide new "gray matter" to store and process the information needed to forage. The whole brain doesn't expand, however; only certain regions enlarge, regions that may provide memory space to store the information about flight and foraging.

This research group and others have also determined that the change from hive duties to foraging is mediated by a hormone called juvenile hormone, which occurs at a low level in the young bee, then increases before the bee becomes a forager. Colony conditions can influence the secretion of this hormone, so the colony can regulate the change from hive worker to forager. For example, if you remove most of the older foraging workers from a colony, the level of juvenile hormone will increase in some of the remaining young workers and they will become foragers within a few days. The next step in this research will be to determine whether the juvenile hormone is responsible for the changes in brain configuration associated with foraging. If so, we will have established a direct link between colony conditions, hormone secretion, and brain structure, and will be able to explain not only the mechanism by which a bee "knows" when to become a forager, but also how it happens.

Such functional linkages are only a small part of understanding what it means to be a bee. What a bee knows, and whether, indeed, a bee can "know" things at all, was another major topic at the Sorbonne meetings. An entire day was devoted to a symposium on cognition in social insects, mostly bees, and each speaker gave his or her views on whether bees have knowledge and can think. Scientist after scientist got up and described how they recorded electrical potentials from bee nerves going to and from the brain, put worker bees into the bee equivalent of rat mazes and watched them try to reason their way out, trained tethered bees to extend their tongues following various signals, and painstakingly dissected and mapped bee brains and nerves after applying various dyes and labels that highlighted active parts of the brain.

My mind kept drifting during the talks, connecting to the ghosts of scholars past that filled the room, and I thought I heard the great philosopher Descartes deliver his famous line about whether humans really exist: "I think, therefore I am." I imagined Descartes and his pupils centuries ago trying to determine whether bees "know" what they're doing, whether an individual bee or colony has any understanding of thinking and existing. The work I was half-listening to in my dreamy daze didn't seem to be providing any answers, nor did the research on bee brains and hormones tell me whether or not cognition exists in bees. Even my Cartesian hallucinations failed me; the abstract and imagined ramblings of bewigged and perfumed philosophers centuries ago didn't seem particularly relevant to bee thought either.

I did, though, begin thinking about an experience all beekeepers have had but may be hesitant to talk about because it might seem flaky. All beekeepers have at one time or another felt "tuned in" to their bees, as if they were inside the colony feeling what it was like to be a bee. Close your eyes, imagine a bright, sunny midsummer day when you have the time and inclination to go through a colony purely to see what the bees are doing. Now relax, take a few deep breaths, and begin to feel the hum of a properly working nest, the bees walking over your bare hands as if they were comb, the sticky feel and smell of honey and propolis, and the underlying feeling that all is right with the hive, that you and the bees know your jobs and are focused on what needs to be done, in harmony with the other bees.

Now come back to reality and think about what you've just experienced, and think about bee brains and bee knowledge. To me, it's a no-brainer; yes, bees "know" things, and understand them from a bee's point of view. The incredibly technical research being conducted today on bee brains and hormones is fascinating because it provides mechanisms by which we can take bee behavior apart and learn how an individual bee determines what needs to be done and how to go about it. It does not, however, and never will, take us that final step to understanding what it is to be a bee and what a bee knows and feels. Science is enormously interesting because it tells us how things work, but the underlying meaning of things cannot be presented at a scientific meeting. Rather, life as a bee sees it can best be felt after spending a few hours in the beeyard on a sunny day. It is ironic that cognition, the most intellectual of topics, is most easily understood by a nonintellectual approach, by feel and touch, by drifting away from the details and techniques that make up science and tuning in to the world from the bee's perspective, a world we are only beginning to understand.

6. Division of Labor

HOW DOES A WORKER BEE KNOW WHAT TO DO when she gets up in the morning? I organize each day with the help of a calendar, a never-ending "to-do" list, and a secretary, but as far as I know, bees can't read or write, and "secretary" is not a task I've seen described in any of the bee literature. A bee's work is further complicated because each day's required work is different. Sometimes a bee might work on a project, such as foraging on blueberry flowers, for many days in a row, while on other days an emergency such as the sudden death of the colony's queen may erupt, and the entire workforce must rapidly shift its workload to respond. Context is everything to a bee, and the success or failure of workers to respond to the changing contexts of colony work determines whether the colony thrives or dies.

A worker bee's ability to respond to a range of unpredictable and complicated conditions is limited, however, by the small size of its brain and the complex and difficult nature of most tasks in the colony. Consider a job like feeding brood, for example. A worker bee must know a few days ahead of time that she will be doing this task because it takes a few days for her to produce and store brood food in her glands. Then she has to find the brood, inspect individual cells to discover whether the larva in each needs to be fed, determine the age and type of the larva so that it gets fed the proper blend of food, and in a few days "decide" to stop producing brood food and move on to another task. Further, each job takes place within the context of the colony, in which the worker is continuously bombarded by stimuli that could distract her from the job at hand. Many of those stimuli are important, and the brood-feeding worker must evaluate the information bombarding her little brain and decide if conditions in the colony require her to change jobs prematurely.

The only way out of this dilemma is for the bee to have simple underlying rules that allow her to move in an orderly fashion through tasks during her lifetime but provide enough flexibility to allow her to depart from the pattern if unusual circumstances warrant it. The solution to the problem is division of labor. Each bee follows a fairly predictable progression of tasks during its lifetime, but the progression can be influenced by various signals in the colony so that bees can shift jobs when necessary.

The task progression followed by workers is based on age; young workers perform tasks inside the nest and older workers guard and forage. Typically, a newly emerged worker spends a few days cleaning cells and capping brood while her brood food glands develop. Then she tends brood and attends the queen for about a week. This is followed by a period during which her brood food glands shrink and her wax-producing glands enlarge. Middle-age workers build comb, handle nectar and pollen within the nest, and do some cleaning. The final phases of the worker's life involve guarding the nest entrance and, finally, foraging. This orderly progression of jobs is mediated by hormones, especially juvenile hormone. The level of this hormone is low in young bees and gradually increases as bees age and move outdoors. Thus, bees have a system of age-based tasks that seems to provide the simple underlying rules that tell them what to do at various stages in their adult lives.

But wait. Suppose there is little or no brood in the colony; should a worker bee produce brood food and neurotically feed empty cells simply because she is at the brood-feeding stage? Or conversely, perhaps there's a baby boom and the colony needs both brood feeders and pollen collectors; how does a bee decide what to do? What happens if the colony suddenly discovers a nectar bonanza in the field next door; who is going to handle the influx of rich food into the nest? Suppose a colony in a nearby tree finds the nest and starts to rob the colony's honey; will some of the comb-building bees walk off the construction site and defend the colony's entrance? The age-based task system alone is not enough to explain how bees allocate jobs. Workers also need to be able to respond to cues in the nest that tell them a shift from the typical progression of work is necessary.

We know a great deal about workers' ability to change the frequency of the tasks they perform and the age at which they do particular jobs. For example, we know that if we add brood to colonies, foraging workers will shift from nectar collecting to pollen collecting, and individual pollen collectors will bring back larger pollen loads to meet the increased demand for protein that the extra brood requires. Removing comb from colonies produces even more drastic effects: some workers will begin foraging at a younger age for nectar to provide fuel to produce more comb, while others will increase their wax production. Conversely, the foraging age can increase in some circumstances. Young workers in colonies newly founded from packages or swarms will delay foraging until after the first brood emerges so that the colony will have foragers available for the second brood cycle.

Although we now understand that such shifts in frequency and age of task performance occur, the current research challenge is to determine which stimuli in the nest induce workers to shift their workloads. This type of research may have considerable importance for bee management because an increased ability to manipulate workers' tasks would enable beekeepers to force colonies to work in ways that might enhance colony productivity. For example, a pollination unit that focused its foraging on pollen would do a better job of pollinating crops, and should therefore yield a higher rental fee to the beekeeper. Another example is package bee production; tricking a colony to rear too much brood a month or so before shaking bees into packages for sale would increase the number of workers that could be shaken from a colony, again yielding more income to the beekeeper. Remember that the stimuli that induce bees to shift tasks must be simple enough for the bees to understand—and thus should also be easy enough for research scientists to discover!

Pheromones, chemicals produced and released by the bees that affect the colony's behavior, physiology, or both, may provide new management tools that influence bee labor in economically important ways. This does not mean that other aspects of colony life don't influence worker tasks. On the contrary, much work in the colony is determined by behaviors that are not based on pheromones. Food exchanging between workers is a good example of a non-pheromone-mediated behavior that influences worker tasks. "Requests" by many workers for protein might stimulate more foraging for pollen, while an increase in incoming nectar influences within-colony workers to receive nectar and store it. These types of task-influencing mechanisms are difficult to manipulate in economically useful ways, whereas the application of synthetic pheromones to colonies as a management technique might supplement the types of colony manipulations beekeepers already perform.

A number of identified and as-yet-unidentified bee pheromones influence worker behavior within and outside the nest. One of the most important is the queen mandibular pheromone, a substance produced in the queen's mandibular glands and spread through the nest by workers that attend her. This pheromone has been produced synthetically and is fairly inexpensive and easy to apply. When sprayed on a colony, mandibular pheromone can delay the age at which workers begin to forage, and in some situations may induce workers to forage for pollen rather than nectar. In packages of bulk worker bees, the pheromone can replace the queen, so that bulk bees can be shipped in a queenless state. Outside the colony, mandibular pheromone sprayed on blooming crops attracts workers, induces them to visit more flowers, and results in increased recruitment by returning foragers of bees in the hive to the sprayed area.

Another pheromone that might be used to manipulate worker behavior is brood pheromone. Odors from brood stimulate the development of the worker's brood food glands, and might be used to regulate the number of workers engaged in brood rearing. Some components of this pheromone were recently identified by French researchers. Once it becomes commercially available, brood pheromone can be applied when brood rearing is a priority for bee management. Thus, a package producer could stimulate worker production by feeding colonies and applying brood pheromone simultaneously. A beekeeper renting colonies for pollination could stimulate brood rearing just before moving units to the crop, thereby providing a colony focused on pollen collection, which would be a better pollinating unit. Finally, brood production could be stimulated in regions where the season is short and rapid colony growth is at a premium, such as the northern areas of Canada.

Division of labor has been studied for some time without providing any obvious advantages for beekeepers, but these basic studies may eventually lead to practical applications. We now know enough about the division of labor to begin custom-designing colonies for particular management purposes so that we can artificially focus a colony's work. Colonies could be designed to focus on pollen collecting, brood rearing, or unusually rapid growth by manipulating conditions within the hive and possibly by providing specific pheromone stimuli. In the end, a custom-designed colony could provide greater income to the beekeeper. One challenge for researchers today is to use this basic information about division of labor to "construct" better colonies and do economic analyses on these improved units to determine whether they are, indeed, moneymakers. After all, if a simple-minded bee can figure out what to do in the nest, complex-minded humans should be able to trick them into doing work our way rather than theirs.

7. Bee Metaphysics and Mr. Spock

BEEKEEPERS, DO PEOPLE SEND YOU THINGS ABOUT bees all the time? I can't begin to count all the cartoons, photographs, mugs, honey jars, toys, stuffed animal bees, honey samples, stickers, ashtrays, patches, and jokes I've been given over the years that have something to do with bees or honey. My office is festooned with this memorabilia, from the large drawing by a friend that says "Comb Sweet Comb" to the stuffed animal and the mug my daughter and my wife gave me one Valentine's Day, each with "Bee My Honey" written prominently on it.

Most of these items are cutesy, jokey things, but there is a more serious side to people's fascination with bees that moves into the areas of philosophy, religion, metaphysics, and even New Age flaky. A friend recently sent me an interesting photocopy of the "bee" entry from a book titled *A Dictionary of Symbols*, by Jean Chevalier and Alain Gheerbrant, which reminded me of this other stream of thought about bees. The bee section begins as follows:

Numbers, organization, unwearied toil and discipline would all make the bee no more than another ANT—the symbol of the masses doomed to endure their fate—were it not that it has wings and a song and distils immortal HONEY from the delicate scent of flowers. This is enough to add a powerful spiritual dimension to the bee's purely material symbolism. Working in their HIVE, a home buzzing with activity and which is naturally equated with the airiness of the artist's studio rather than with the gloom of the factory, bees collectively ensure the survival of their species. Yet taken as individuals, a universal quickening power between Heaven and Earth, they come to symbolize the vital principal and to incarnate the soul.

My first reaction was probably the same as yours: Give me a break. I mean, bees as the quickening power between Heaven and Earth, immortal HONEY, delicate scent of flowers, the hive as an artist's studio? When was the last time you attended a bee meeting where a speaker lectured about bees in such a metaphysical way? I'm sure virtually every beekeeper in the room would head for the hallways for a smoke or a coffee break to talk about the latest way of illegally using miticides until this flaky speaker had finished connecting bees to the eternal cosmos.

That might be the reaction of those of us used to thinking about bees in a functional way, focused on their biology or management rather than their part in the grander universe. To many people, however, bees have enormous spiritual, religious, and philosophical importance, and both bees and honey have been a part of human lives and thoughts for millennia. Indeed, there is a long tradition that considers bees for their spirituality rather than their utility, and bees have appeared in the great books and writings of virtually every major religion.

Take Islam, for example. Muslims say true believers are like bees that have chosen the fairest flowers to visit for nectar; in one Muslim tradition bees are considered to be angels. The ancient Egyptians believed bees were born from the tears of the sun god Ra falling to earth, and the great Greek philosopher Plato thought bees to be the souls of the righteous reincarnated. Even mainstream Christianity has gotten into the spiritual side of bees. Medieval Christians thought the buzz of the bees was a song that embodied a spark of the divine, and considered the three-month winter season when the bees don't appear outside the hive as representing the three days after Christ's crucifixion when his body vanished.

In addition to symbolizing the spiritual side of things, bees have also have been used to represent eloquence, poetry, and the mind. It was thought that Saint Ambrose's eloquent sermons and musical ability were the result of bees touching his lips while he was in the cradle. Indian writers waxed eloquent comparing bees visiting flowers to the soul sucking on the intoxicating pollen of knowledge. In Hebrew, the term for *bee* and that for *word* come from the same root, and bees and honey are mentioned innumerable times in the Old Testament.

People unfamiliar with bees have a fascination with them that goes beyond simple interest in another species. My city has a science museum with a glass-walled observation hive, which invariably is surrounded by a huge crowd of children and adults. Some of them are doing something specific, such as trying to find the queen or see a bee dancing, but others seem transfixed by the hive, absorbed in its activity without trying to find anything in particular. They look and act as if they were meditating, and if jerked back to the present by a question or bumped by a neighbor, they seem confused for a minute, as if they had forgotten where they were.

I've seen this same reaction in students unfamiliar with bees who view the inside of a hive. It seems to happen the second or third time they work bees rather than the first, which is taken up by fear of stinging and figuring out the mechanics of veils, smokers, and hive tools. Once they're comfortable, however, a detached focus settles over their faces, the chatter stops, everything seems to slow down, and they become absorbed in the nest. "Cool" is the way they describe it, but I think they mean much more.

I know what they mean because I've felt it, and you probably have, too. Not on those rushed days when you have a few hundred hives to feed and have to get home to take your daughter to ballet or your son to baseball practice (or, if you want to be more politically correct, your son to ballet and your daughter to baseball practice). No, this feeling of connection with the bee world comes only on slower days, perhaps during the honeyflow, when you're going through the occasional hive just to see what's going on. There comes a moment when you pause, and deeper senses than sight, sound, and smell kick in, in some intangible way giving you a feeling that something else is there with you. That "something else" is nothing bizarre, odd, or even alien; it's just a bee colony, but in those moments the bee's point of view takes over from the human perspective.

I can remember when it first happened to me. I had just arrived in South America to begin my Ph.D. research on Africanized bees, and had virtually no experience with honey bees. I approached my first hive, a small nucleus colony, suited up from head to toe for protection, since I had only the bees' reputation as a guide to what might happen. I smoked the colony until I couldn't see the bees for all the haze, and then gingerly went in to take off the top, as if I was disarming a bomb rather than looking at bees.

Of course, nothing remotely frightening happened. The colony was small, and even if the bees had wanted to go after me they were so heavily drugged by the smoke that I doubt they could have found me. As I went through the frames, a strange sensation came over me. To my astonishment I was relaxed, peaceful—the last thing I had expected to feel while working a colony of "killer bees."

This calmness soon expanded to the point that I took off my gloves and veil, and forgot what I was supposed to be looking for, which was probably the queen or something mundane like that. Rather, I felt an attachment to the bees in some way I still can't explain, much as someone might feel when meeting a natural soulmate for the first time. For a few minutes, at least, I thought I knew what life must be like from the bee's point of view, walking around in the hive surrounded by my sisters, smelling the smells, sensing the vibrations, touching our antennae, caring for our young.

I don't click into similarly poetic bee experiences every time I open a hive, of course. There's usually too much to do, too many details to explain to a student, or simply too many heavy supers to lug around, and I don't have the time or focus to slip into gaga land. It does happen, however, and although few of us talk about it, I have a feeling it happens to you, too.

If so, you're part of a long tradition that considers bees worthy of religious experience, and honors this species in imagery and with respect for what bees symbolize to humans. This knowledge is not like learning to feed sugar syrup, or figuring out the right time to super, or finally getting all your disease and pest control down right. This is a way of understanding bees that can't be taught and isn't discussed in *The Hive and the Honey Bee* or *The ABC's and XYZ's of Beekeeping*. It can be experienced only by letting yourself become conscious of the world as bees might see it. It's a bit like Mr. Spock doing a Vulcan mind meld with a worker bee or Deanna Troi having an empathic experience with the queen. Perhaps the hippies of my generation expressed this connection with bees best in their classic 1960s comment on the world in general: "Like, wow."

PART TWO
In Sickness and in Health

In Sickness and in Health The art

of beekeeping is no different from any other craft. It is clear in concept, deceptively easy to learn, and enormously complex and subtle for those who probe its depths. Contemporary beekeeping is still simple in theory, but it has become more complex in practice because the bee diseases and parasites beekeepers have to deal with have increased in number and severity. The craft of bee research is also deceptively simple, especially where management research is concerned. The elegant and clearly phrased hypothesis and the well-designed experiment remain at the core of bee science, but the techniques, issues, and approaches of management research have grown in complexity as bee management itself has become more intricate.

The basics of beekeeping can be explained in one sentence: A colony should be managed so that its population peaks coincidentally with the honeyflow, the time when the maximum amount of nectar is available in the field. Managing colonies to that ideal involves providing them with a young, fecund queen; feeding them sugar or pollen in the spring to stimulate growth; preventing swarming or colony reproduction by providing sufficient space for the growing population; and controlling diseases and parasites throughout the year.

Would that it were so easy. Take nutrition, for example. Under natural conditions bees need only nectar and pollen as food: nectar to supply carbohydrates, and pollen as a source of protein, fats, minerals, and vitamins. But managed colonies must be fed sugar to replace the honey beekeepers harvest, and a protein supplement to stimulate and fuel brood rearing. The sugar part is easy; bees do well when fed simple table sugar. Protein is more difficult; bee scientists have failed to invent a protein-based food substitute for natural pollen, in spite of many decades of work and hundreds of studies. Supplements based on brewers yeast, soy flour, fish meal, and milk by-products can in part replace pollen; however, if used alone these supplemental foods are effective only for short periods. Worker bees soon begin to exhibit stunted growth, unusual behaviors, and lethargy, and colonies fed only supplements for more than a few weeks dwindle and die.

Or take the queen as another example. It is a simple matter to purchase a new queen from a beekeeper who specializes in breeding and rearing queens for sale and insert her into a colony to replace an old queen. It is not so easy to find a prolific queen whose worker offspring combine disease resistance with the gentle demeanor and work ethic needed to produce copious quantities of honey. The bee's genetic system works against the breeder. The female queens and workers have the usual two sets of chromosomes, but the male drones have only one set; thus, inbreeding is an occupational hazard for breeders attempting to produce and maintain large numbers of queen bees for sale.

The queen-breeding industry is further complicated by a beekeeping fact of life: most beekeepers want new queens in the spring so they can start new colonies or split established overwintered colonies early enough in the season for those colonies to grow to sufficient size to produce honey during the summer. For that reason, most queen rearers work in the southern United States, where queen rearing can begin as early as January or February. That was not a problem until Africanized bees from Latin America spread into Texas, Arizona, New Mexico, and California and limited the area in the southern United States where queen rearing can easily be accomplished. Feral Africanized drones can mate with queens from beekeepers' colonies that are being reared for sale, creating queens whose hybrid offspring express undesirable Africanized traits, especially the famed defensive behavior. The continued spread of Africanized bees into the southern queen-rearing states threatens the supply of desirable queens for all U.S. beekeepers.

Another problem facing North American beekeepers is importing new stock. Beekeepers face the same problem that farmers growing wheat, corn, cotton, apples, or any other crop must deal with. Most of the crops grown commercially in the United States and Canada originated elsewhere in the world, and bees are no exception. The honey bee evolved in Africa and Europe, and was brought to North America from Europe by the first settlers. The importation of bees and agricultural crops created a limited gene pool with relatively little genetic variation with which to respond to diseases and parasites. The obvious solution, importing new genetic stock to increase variability, is somewhat problematic because imported stock often brings in new pests along with new genes.

Agricultural regulators are aware of these dangers, and it has been illegal to import bees from offshore into the United States since the 1920s. Unfortunately, this ban did not prevent individual beekeepers from putting their perceived need for new stock ahead of the law. Beekeepers afflicted with a "grass is greener on the other side" perspective smuggled in queen bees that carried the *Varroa* and tracheal mites that have devastated the North American beekeeping industry over the last fifteen years.

The tracheal mite first appeared as a pest in England early this century and wiped out a considerable portion of the British beekeeping industry. This mite lives and reproduces inside the breathing tubes of bees, the trachea, thus its common name. Similar, generally harmless, mites are found externally on most bee species, and the tracheal mite is thought to have evolved from one of these external mites that developed the habit of living inside the honey bee and feeding on the bee's blood through tiny holes drilled in the lining of the bee's trachea.

The *Varroa* mite came from Asia, where it is a minor pest on *Apis cerana*, an Asian honey bee closely related to the honey bees imported into North America from Europe (*Apis mellifera*). Like many parasites, *Varroa* mites, which have no common name, can parasitize closely related host species that have few defenses against them. The Asian bees have evolved grooming mechanisms that eliminate most of the mites from their nests, but European bees have no such grooming behavior, and the mites thrive in their defenseless colonies. The mites swept through the North American beekeeping industry with the speed and severity of a hurricane, and most of our hives would die within a year or two of infestation without treatment.

Beekeepers have responded to both mites with the classic pest management techniques of developing genetically resistant stock and using chemical pesticides for control. To date, breeding in genetic resistance has been only marginally effective, and it is chemicals that are keeping beekeeping alive. Beekeepers and bee scientists have mounted Herculean efforts to select for mite-resistant stock, sometimes importing queens and drones from Europe into intensive quarantine situations and sometimes attempting selection from indigenous North American stock. Breeders have had some success at breeding resistance to tracheal mites into bees, but the colonies headed by mite-resistant queens often have other attributes that are not desirable for hobby and commercial beekeeping. There is little evidence that the tracheal mite–resistant stock is resistant enough to completely eliminate the need to use pesticides within colonies, although resistant stock may reduce the need for them. Genetic resistance to *Varroa* is still only on the horizon. To date, there are no commercially available queens with the level of resistance necessary to combat *Varroa*.

That leaves chemical control, a decidedly unpopular solution in the view of pesticide-phobic beekeepers. Chemical control of tracheal mites is less of a concern to beekeepers because tracheal mites can be kept at low populations by such relatively benign substances as vegetable oil, menthol, and formic acid. The vegetable oil, which is mixed up in a hamburger-shaped patty using pollen supplements as a base, seems to coat the bees with a thin layer of oil that confuses mites trying to move from one bee to another. Menthol and formic acid are put into hives as fumigants and probably work by drying out the mites, which are more susceptible to desiccation than bees because they are smaller. Vegetable oil and menthol are common additives to our own foods, and are largely unregulated substances. Formic acid is a natural component of honey, and low-level residues of formic acid are not considered a problem from a consumer's perspective. However, formic acid poses some danger to the beekeeper if proper application procedures are not followed. Its use on bees is regulated in Canada and is not permitted in the United States, although federal regulators may soon relax the ban.

Varroa control is much more problematic because it involves using a harsher pesticide, fluvalinate, which can leave unacceptable residues in honey and wax. Beekeepers both hate putting fluvalinate into their hives and are concerned that the mites will develop resistance to this chemical, as has been the case in Europe, which would leave U.S. beekeepers with no registered chemicals to use against the *Varroa* mite.

The problems caused by mites have not been limited to the beekeeping industry. Honey bees pollinate some ten billion dollars worth of crops in North America: apples, pears, cherries, almonds, oilseed, raspberries, blueberries, melons, squashes, and a great many more. This heavily managed system works because beekeepers truck their colonies from blooming crop to blooming crop throughout the season, receiving a small fee from growers for providing the pollinators. However, managed pollination requires more than a million colonies in the United States each year, and these colonies need to be strong and healthy in the spring. The weakened and dead colonies resulting from mite infestations have put a considerable strain on the ability of beekeepers to provide sufficient colonies for pollination, which has created interest in developing pollination management systems using other bee species that are not susceptible to the mites.

The parasites have created an interesting environment for the research community, and have also stimulated a closer relationship between beekeepers and researchers. Beekeepers have had to learn how scientists work and have come to appreciate the rigor that must go into proving the effectiveness of new control methods. Many beekeepers have been performing their own research, however, frustrated by the seemingly slow pace of scientists, and in some cases have spread unverified and downright incorrect information about silver-bullet control methods around the beekeeping world via the Internet. Scientists and extension agents have found this enormously frustrating, since poorly documented snake oil cures have taken on the aura of "fact" just because they appear repeatedly at different Web sites and in unreviewed beekeeping newsletters.

Scientists have had to learn how beekeepers think and work, too, and perhaps this has been one of the hidden benefits of killer bees and mites. I may be imagining it, but it seems that more of us are working in partnership with beekeepers to design and perform trials of novel mite controls, and certainly all of us are working with bee-keepers to generate the financial resources necessary to look for and examine new methods. In the end, beekeeping may change less than bee research. Beekeeping will always be an industry in which hard work and simple management can lead to success, even if our management repertoire needs new tools. Research, however, is changing. It is still a profession based on hard work and simple, careful design, but bee researchers are being forced to tackle the immediate problems facing the beekeeping industry. To my thinking, that may be a really healthy result of bee disease and Africanized bees.

8. Hybrid Bees

I'M SURE IT'S NO ACCIDENT THAT THE WORDS *EN-tomology* and *etymology* are so similar to each other, and so easily confused. Entomology, of course, is the study of insects, and etymology is the study of the origin of words. Entomologists are great wordsmiths, and our meetings percolate with long, descriptive terms about the insects we study. Heaven for entomologists is discovering a new insect species and getting to name it, because we can then search dictionaries and our memories for Latin words that in some way capture an essential quality of the new insect. One such name was given to the honey bee centuries ago: *Apis mellifera*, which means "honey-bearing" or "honey-producing bee" in Latin.

Thus, it's not surprising that the first thing I did before writing about hybrid bees was look up the term *hybrid* in the dictionary. The current definition is "offspring of the union of a male of one race or variety with the female of another," or "derived from unlike sources." However, the Latin origin of the word *hybrid* is *hybrida*, which has a much more descriptive definition: "the offspring of a tame sow and a wild boar." The more poetic Latin version captures the essence of the hybrid dilemma for queen breeders: how to mix two varieties of bee in such a way that the offspring queen has the best characteristics of the tame sow and the wild boar rather than the worst of each. That is, the desired outcome is a hybrid exhibiting characteristics of the tame bee, but with some of the "vigor" associated with the wild one, not a wild bee whose aggressive characteristics overwhelm the beneficial traits of the tame one.

The positive aspects of hybrid bees have given them a good reputation among U.S. and Canadian beekeepers. For example, one of North America's foremost queen-rearing companies advertises its bees as being "the result of crossing 3–5 different lines, increasing hybrid vigor," and "having the advantage of hybrid vigor." Their premier line of queens is a double hybrid, "a combination of the best of both lines." The term *hybrid vigor* is a valid scientific concept. It describes a phenomenon in which the hybrid organism is superior to both of the original types. The meaning of *vigor* is difficult to pin down, and the genetic basis for hybrid vigor is not well explained. Nevertheless, "vigor" is a good description of what can happen when two unlike types are mixed. Whatever may be behind hybrid vigor, the continued popularity of hybrid bees among beekeepers is evidence that it is a useful concept in bee breeding.

A good, tangible example of how hybridization can produce bees that are commercially superior to their parent stock is mite resistance. A number of studies have shown that interbreeding two lines of non-resistant bees produces hybrids with improved resistance to mites. This phenomenon appears to result from combining traits of the parents in new ways to yield an improved offspring.

Hybrid vigor can be a double-edged sword, however. The most significant type of hybridization taking place today is between African and European bees, and it is yielding an Africanized hybrid that has created numerous real as well as political problems for beekeepers throughout the New World. The Africanized bee originated from an attempt to capitalize on the concept of hybrid vigor by mixing the traits of bees from Europe and Africa. The plan was to produce a hybrid bee that would be more "vigorous" in tropical regions. Honey bees are not native to either North or South America, and up until the 1950s the bees being used throughout the New World came from Europe. These European bees, although fairly gentle, are not good honey producers in tropical climates, and are particularly unsatisfactory in the Amazon Basin. Brazilian geneticists reasoned that they could import more aggressive bees from Africa with better honey-producing characteristics, breed them with the European bees, and end up with a gentle bee that was a good tropical honey producer.

The rest, as they say, is history, and we now are dealing with a number of interesting problems that these hybrid bees have created. The original African type is still predominant in most of Latin America, and it is interesting that there has been so little hybridization of African and European bees in feral colonies throughout the tropical regions of the New World. Beekeepers on both New World continents are attempting to maintain European colonies, but often end up with Africanized hybrid bees through mating. Unfortunately, the hybrid bees seem to maintain more of the African traits than the European ones. That is, the characteristics of the wild boar, the African drone from feral colonies, overwhelm those of the tame sow, a virgin European queen in a managed colony.

One very obvious problem caused by the hybrid bees is their pronounced defensive behavior at the hive. The increased defensiveness of Africanized bees comes from two sources: the individual bees themselves and the mixed colony that contains both Africanized and European bees. Feistiness is a genetically dominant trait in the individual bee, so the offspring of a European queen and an African drone will maintain the aggressive traits of the African father rather than the recessive gentleness of the European mother.

The colony provides an additional level of hybridization, which has been studied by the team of Ernesto Guzman (who now works for the Mexican government) and Rob Page (from the University of California at Davis). Since queens mate with a number of drones, a colony may be a mixture of European (European drone mated with a European queen) and Africanized (Africanized drone mated with a European queen) workers. Unfortunately, even a minority of Africanized bees in a colony will dominate behaviorally, and a colony with only 25 percent Africanized workers will behave more like an aggressive African colony than a gentle European one. The European workers appear to have poorly developed self-images and, much like inner-city gang members, are easily swayed to misbehave by a few aggressive leaders. In psychological terms, the Africanized bees are "inner-directed," following their own genetic background, while the European bees are "outer-directed," taking on the traits of the bees around them rather than staying true to their own gentle nature.

In addition to being difficult to handle, Africanized hybrids may be creating a real political mess in North American beekeeping. Many states are developing certification programs that require bees to be tested for Africanization, and Canada is likely to prohibit the importation of bees from the mainland United States without such certification. Although the European bees themselves seem to do a good job of detecting the presence of Africanized workers in their colonies, our scientific detection methods are not as reliable. Again, the team of Guzman and Page, with the addition of their ace queen bee inseminator, Kim Fondrk, have recently shown that the Africanized hybrid bees are difficult to detect with the currently used methodology.

There are many methods of certification, but the most common and cheapest to date involves measuring the body parts of bees. There is a short-cut method that uses primarily wing measurements, and a longer, more expensive system that uses up to twenty-six different body measurements to confirm the short-cut diagnosis. The other methods employ various types of chemical identification, including cuticle type, proteins, and DNA analysis. These methods are more reliable, but they are also expensive and technically difficult to conduct. The Guzman study found that the morphological method was accurate for detecting pure African and European bees, but hybrid bees were frequently misdiagnosed; in fact, up to 50 percent of the hybrids were classified in the wrong categories. This is a serious problem for certification because most "problem" bees would be hybrids rather than pure African. Since the hybrid bees tend to be as aggressive as pure African bees, misdiagnoses could create serious problems for beekeepers who depend on an accurate certification procedure, and for a public that may not tolerate mistakes in identification.

Guzman and Page made another interesting discovery about Africanized bee detection. In one sense, it's not all that important to diagnose individual bees as Africanized or not; it is a colony's behavior, not its genetics, that we need to examine. Would a gentle Africanized bee cause a problem, or should we simply be screening against the extreme defensive behavior that some colonies exhibit, no matter what their genetic origin? Unfortunately, it is impossible to screen the defensive behavior of all the workers before moving a colony, leaving us with the possibly unreliable method of morphometrics or the overly expensive chemical methods as certification techniques.

These findings do not bode well for those interested in freely moving bees in the United States or exporting them to Canada. It is unlikely that Canada will open its borders to U.S. bees without inexpensive and unambiguous certification methods to determine the extent of Africanization, and at the moment these do not exist. Guzman's work will undoubtedly come under considerable scrutiny over the next few years because of its significant economic implications, not only for Canadian importations but also for required certification purposes within the United States.

The arcane tame sow and wild boar of ancient Latin origin turns out to have a very contemporary relevance for today's beekeeping, and the term *hybrid* may no longer have the same positive connotations for commercial beekeepers it once had. We're undoubtedly going to hear much more about hybrid bees in the future. Buckle your seat belts; it's going to be an exciting entomological and etymological ride!

9. Let's Do Lunch

I HAD A DREAM THE OTHER NIGHT ABOUT A NEW type of beekeeper, the Yuppie Beekeeper. In my dream, a thirty-ish guy in a three-piece suit and spit-shined Gucci dress shoes showed up at an apple orchard, carrying a fancy leather briefcase in one hand and a hive tool in the other. It was spring and the apples were in full bloom, the air fragrant with blossoms and the promise of a bumper fall crop. The Yuppie Beekeeper approached the owner of the orchard, who was also dressed like a Wall Street banker and was driving his tractor through the trees. The grower hopped off his tractor at the beekeeper's approach with a great big grin on his face, shook hands, and said, "You were right. Pollination is a win-win situation for both of us. Let's do lunch."

Fortunately, I woke up before I had to dream through a yuppie lunch, but I was puzzled by my dream. After all, with all the problems and hard times in agriculture, and the grousing we hear from farmers, how could a beekeeper and a grower have struck it rich? Well, in real life our beekeeper and his pollination client would not have turned into yuppies, and probably would not have struck it rich, either, but good pollination management can increase the incomes of both parties. Indeed, there are a number of trends in beekeeping research and management practices that point to a diversified pollination management system as one of the best ways to increase the beekeeper's income and improve the grower's crop yields and income as well.

The single most significant trend affecting commercial pollination today is a shortage of bee colonies. Tracheal and *Varroa* mites have taken their toll, and growers from the almond orchards of California to the blueberry fields of Maine are reporting considerable difficulty obtaining sufficient bees to pollinate their crops. Couple that with impending movement limitations that may result from the spread of Africanized bees, and there's a real problem. The solution, however, can involve more efficient management of both honey bees and other pollinators using methods that in the end may provide more profit to beekeepers who provide pollination services to growers.

What I propose is simple. The research community needs to document how certain improvements in pollination management can increase bee visits to flowers, thereby improving fruit set, crop yields, and, ultimately, profits to growers. Beekeepers need to be willing to implement new techniques that both intensify and diversify their management systems. The bottom line, of course, will be higher pollination fees paid to beekeepers and greater profits to growers due to yield increases. Let's look at some examples of how pollination research can be transformed into more dollars for both parties.

An excellent way for beekeepers to increase their pollination income is to provide better colonies to growers, which allows them to increase the pollination fee they charge per colony. By "better" I don't necessarily mean bigger, but rather colonies that are more focused on pollen collection. One of the few things we do know about pollination is that bees collecting pollen are better pollinators than those collecting nectar, because pollen-collecting bees do a better job of moving pollen between flowers, resulting in better fertilization of the plant's seeds. Ultimately, this improved pollination produces higher yields of larger and more uniformly shaped fruit.

There are a number of methods beekeepers can use to induce workers to forage for pollen rather than nectar. A very effective one is to manipulate colonies so that the units used for pollination are full of young brood but have little stored pollen. The simple procedure of removing frames of stored pollen before moving colonies to crops can increase pollen foraging by 10 to 30 percent, which should result in better crop pollination and higher yields. A highly effective method used in New Zealand kiwifruit pollination is to feed sugar syrup while the bees are in the kiwifruit orchards. This technique can double pollen collection, presumably because colonies perceive their need for pollen to be relatively greater than their need for nectar. The result: win-win. Beekeepers can charge more for these improved colonies, and growers can afford to pay it because their higher yields have increased their income.

Another technique that improves pollination is to spray attractants on the crop during bloom. We have been working extensively with one such attractant, called "Fruit Boost," that is composed of a synthetic blend of honey bee queen pheromone. These compounds are highly effective at attracting bees to crops, and can increase yield characteristics such as fruit set and size by 5 to 30 percent or more through improved pollination. Fruit shape may be more uniform as well, increasing the number of fruits in the higher grade categories, which produce the best profit for the grower. The cost of the spray is only about thirty dollars per acre, but net profits show average increases of four to five hundred dollars per acre on such crops as blueberry and pear. Beekeepers providing colonies to growers might wish to provide a more complete pollination "service" that would include spraying attractants to increase the numbers of bees foraging directly on the target crop. Again, beekeepers can charge more for this service, and growers should be willing to pay because they make more profit.

A third way beekeepers can diversify their pollination service is to provide other pollinating species such as leaf cutter bees or bumble bees to supplement honey bee colonies. Leaf cutter bees have been used for some time to pollinate alfalfa for seed production, because they work the alfalfa flower more effectively than honey bees. Bumble bees are possibly the best general pollinators of all the bees; they move rapidly between flowers, carry large pollen loads, and buzz flowers when they visit, thereby dislodging copious amounts of pollen. Until recently, however, bumble bees were not an option for pollination management because they were too difficult and expensive to rear.

New methods of bumble bee culture developed since 1990 have made it commercially viable to use these bees for pollination management. Today, thousands of bumble bee colonies are produced in huge warehouses, primarily for the greenhouse market. Although a single colony can cost many hundreds of dollars, their use in greenhouses is highly cost-effective when compared with the expense of the hand pollination methods previously used for such high-value greenhouse crops as tomatoes and peppers. Honey bees, in contrast, do not orient and pollinate well under glass, and so have not been used extensively for greenhouse pollination.

Bumble bee management has good potential for field crops as well. The commercial availability of large numbers of bumble bee colonies has stimulated research into their use on berries and fruit crops, with some early success. Bumble bees do a better job per bee of pollinating many crops, particularly in areas with cool, wet weather or a shortage of honey bees. The negative side of bumble bee pollination is the high cost per colony, although larger-scale rearing methods and techniques to utilize colonies for three or more pollination sets will eventually bring down the cost.

Thus, today's beekeepers can offer a diversified pollination service to growers by providing more intensively managed honey bee colonies, using attractant sprays, and perhaps even supplying two or more species of bees to tailor pollination services to the needs and wishes of the grower. In return, of course, beekeepers should charge a considerably higher fee.

The key "missing link" in selling diversified pollination management to growers is proving that the higher pollination fees are cost-effective. Indeed, economic analysis of innovative bee management techniques is an aspect of bee research that generally has been neglected by the research community. Applied research often does not take the final step; researchers do a good job of determining the *potential* of a new idea, but don't often follow through to show the costs and benefits of the new techniques. Yield increases are the bottom line for growers, and this is the area researchers need to examine if we are to justify cost increases to growers for improved pollination services.

Are yield increases due to more intensive pollination management a realistic expectation? I think so. Only small increases in yield are needed to justify higher fees. Approximately $10 billion worth of crops are pollinated by bees in the United States and Canada. If a 1 percent average yield increase resulted from better pollination, the crop value would improve by $100 million. Surely beekeepers would be justified in asking for $10 million of that in increased fees. Further, a 1 percent increase is a very conservative estimate; attractant pheromone sprays alone can increase crop value up to 30 percent in some cases, with average increases of 5 to 10 percent. I'm confident research into other management techniques such as feeding colonies and using other bee species to supplement honey bee pollination will also demonstrate yield and profit increases in bee-pollinated crops.

It's time to stop moaning about low pollination fees and the shortage of colonies for pollination. Instead, beekeepers need to prove the significance of good pollination management to growers and then be aggressive about charging appropriate fees for our services. The approaching pollination crisis due to colony shortages can be a real opportunity for beekeepers to demonstrate the value of pollination management to the agricultural community, and to insist that we are paid properly for the necessary and economically significant service we provide.

Let's do lunch.

10. Pesticide Resistance

THE NORTH AMERICAN MEDIA HAVE BEEN HIGH-lighting a new scourge on humanity in the 1990s, flesh-eating bacteria. In Canada, where I live, a prominent politician lost his leg to these hungry consumers, generating a flurry of news reports about this and other diseases that are becoming resistant to antibiotics. Modern medicine has become the victim of its own ingenuity and success, and it seems our drug producers are barely keeping ahead of evolution. Almost as soon as a new designer antibiotic hits the market, new strains of bacteria appear that are resistant to its effects. Old diseases such as tuberculosis are returning to plague us, and our high-tech, supposedly antiseptic hospitals are reporting high mortality rates from people who should survive but are struck down by microbes that were easily controlled a few years ago by the first generation of antibiotics.

Until recently the honey bee world was mercifully spared this problem; none of the treatable bee diseases showed any sign of becoming resistant to beekeepers' antibiotic arsenal. The best example is American foulbrood, or AFB. Despite fears that the continued use of Terramycin would induce a resistant strain of the bacteria that causes this disease, Terramycin continues to be effective in North America. Samples of AFB from contemporary hives are killed just as easily by Terramycin as bacteria kept in the freezer for fifty years, an indication that this disease has not developed noticeable resistance during that time. A 1996 report of resistance developing in Argentina, however, suggests there may be cause for concern on the horizon.

Varroa mites have changed the picture and are a clear example of the speed at which resistance can develop. Studies in Italy in the early 1990s, and in France in the mid-1990s, found that *Varroa* had become resistant to fluvalinate, the pesticide in Apistan strips, after about ten years of use. We expect 95 to 99 percent of *Varroa* to be killed by a proper application of Apistan, but the Italians found only about 80 percent mortality, which left enough *Varroa* to multiply and kill a colony within a few months. First the Italians, and then the French, were forced to switch to more toxic substances that are not, and likely never will be, licensed for use in North America.

What is particularly frightening about this scenario is that no other registered chemicals effective against *Varroa* are available in the United States, so resistance to fluvalinate would quickly devastate the U.S. beekeeping industry. Even more frightening is the fact that many commonly used management practices are perfectly designed to select for resistant *Varroa* mites. If I wanted to do an experiment to produce resistant mites, I would do nothing more than what is being done illegally today by some North American beekeepers, and I could virtually guarantee that within five years or less I would have mites resistant to Apistan.

The development of resistance to miticides is a fairly common and simple phenomenon. Mites feed on a wide variety of plants and animals, which have evolved numerous protective chemicals to defend themselves from the mites. In most cases, the hungry mites then developed enzyme systems able to break down these chemical defenses. This ability of mites to detoxify natural compounds works on artificial compounds as well. Thus, when mites encounter a commercially produced miticide, a few of them will survive because they have the ability to break down the novel chemical. The surviving mites reproduce, and their offspring thrive because the more susceptible competing mites have been killed off, leaving a wide-open field for the resistant mites to take over. Ironically, our usual pest management response to these survivors is to throw ever-higher doses of chemicals at them, which further selects for even more resistant mites.

Apistan is the type of product that can induce resistance even if properly used. Its active ingredient, fluvalinate, kills mites by disrupting their nervous system, and is thought to be more deadly to *Varroa* than to bees simply because the mites are so much smaller; a dose lethal to a mite won't kill a bee. Current recommendations all across North America suggest two or even three applications of Apistan each season, for forty-two to forty-five days at each application. Spring and fall applications are recommended, both so the honey won't be contaminated and because the mites are more exposed to the chemical when they are out of brood cells and on adult bees. Two or three annual treatments are necessary to keep mite levels down, and the consequences of less frequent or no treatments are diminished honey production and high colony mortality.

These repeated applications of the same miticide within a single season follow both label and extension agent recommendations. But even properly applied treatments may select for resistant mites, since the mites that survive the first exposure to Apistan and reproduce are likely to be the ones most able to detoxify the fluvalinate. A second or third application of the same substance a few months later can select for mites with even more resistance. Proper management against *Varroa* should involve alternating chemicals, so that the few mites resistant to one substance are killed later in the season by the other. However, beekeepers in the United States don't have any alternative chemicals, which places them on the horns of a real dilemma: if you don't treat, your colonies will die, but if you treat at the recommended levels you face the possible development of resistant mites.

Shipping queens and packages with Apistan strips or tabs is another "recommended" use of Apistan that can select for resistance. Tabs and strips in packages kill most of the mites, but which mites do you think are most likely to remain alive? Ironically, shipping bees with Apistan could be a very strong selective agent for resistance, because the mites have no refuge from the chemical, and the few that survive will be the ones that deal with the chemical most effectively. Even worse, shipping mites around North America will increase the spread of resistant mites, just as it rapidly increased the original spread of both *Varroa* and tracheal mites.

Although even proper applications of Apistan have the potential to select for resistant mites, it is beekeepers' misuse of Apistan that will select for resistance at approximately the speed of light. For example, some beekeepers keep strips in their hives almost year-round, except during the honeyflow, so the mites are exposed to Apistan for most of the fall, winter, and spring. There seem to be two reasons behind this continuous chemical application. First, putting strips in and taking them out is a lot of work. Second, there seems to be some feeling that if a forty-five day exposure is effective, a two-hundred day exposure will really blast the dickens out of those mites. In reality, long-term exposure to any agricultural chemical is a recipe for rapid evolution of resistance. Miticides are powerful selective agents, and their continuous use will select very, very strongly for mites resistant to their effects. Furthermore, continuous application results in residue buildup in comb, so that exposure to fluvalinate may continue even after strips are removed, and the honey could be contaminated.

Many beekeepers reuse strips to save money, and simply don't believe the company line that Apistan strips are not effective following one application. Most of the data on this subject are company property and have not been released to the scientific and beekeeping community. An independent study conducted in the state of Washington found that only about 20 percent of the fluvalinate remained following an application, but additional studies are needed to verify this conclusion under diverse treatment circumstances.

Why is reuse of strips dangerous in terms of resistance? The continuous application of a chemical at sublethal dosages is as good a way to select for resistant mites as overdosing them with high amounts of the chemical. Generally, good management against the evolution of resistance involves a short application of a moderate chemical dose. Prolonged applications of any dosage, or shorter but repeated applications of high dosages, will favor the survival of resistant mites and generate a problem that none of us wants to deal with.

What can be done to minimize the possibility that resistant mites will evolve in our hives? The best long-term solution is to find alternative chemicals so that we can rotate them and reduce the mites' exposure to these strong selective agents. Two miticides are licensed for use in Canada, Apistan and formic acid, and alternating their use is highly desirable. Formic acid is thought to act as a general caustic or desiccating agent rather than by disrupting a specific aspect of the mite's nervous system, so mites are less likely to develop resistance to formic acid than to Apistan. Formic acid also has the advantage of being effective against tracheal mites as well as *Varroa*. However, formic acid is difficult to apply safely, is not as effective against *Varroa* as Apistan, and must be put on four or five times at weekly intervals in order to work at economically effective levels. Research is continuing with formic acid to provide a more effective, slow-release formulation that only needs a single application, and an improved release system should solve these problems.

The development of alternatives to Apistan may be the single most important research priority in North American beekeeping today. Until formic acid is made more usable or a new effective miticide is available, however, beekeepers need to exert all possible pressure on their peers to use Apistan only according to label directions. Proper use of Apistan may eventually lead to resistant mites, but improper use is guaranteed to unleash a strain of *Varroa* that none of us wants to have to deal with. Resistant *Varroa* mites would be much like flesh-eating bacteria; they would consume our hives almost as we watch, and there would be nothing we could do about it.

11. *Billions of Pounds*

A BILLION IS A BIG NUMBER, A VERY BIG NUMBER. There are only a few things on earth that we measure in the billions, including the total human population, the national debt, and the net worth of a very small number of billionaires. Even honey production is not measured in the billions. In 1996, for example, U.S. beekeepers produced only 198 million pounds of honey, less than a pound for each American citizen, and considerably less than a billion. Beekeepers' colonies did pollinate about ten billion dollars worth of crops in North America that year, however, putting us in at least one billionaire category and providing some bragging rights at the exclusive billionaires' club.

Beekeepers today are participating in another billionaire category, however, one we shouldn't be quite so proud of. In 1993, 1.1 billion pounds of active pesticide ingredients were used in the United States, or about 4 pounds of pesticides for every man, woman, and child. Considering that most pesticides are toxic to humans in doses of about one hundred thousandth to one millionth of a pound, that's a lot of poison. Even if pesticides were not poisons, but simply dust, that would be a lot of material. A billion pounds of dust dumped on the United States each year would be alarming; a billion pounds of poisonous dust seems catastrophic indeed.

About three-fourths of that poison was used in agriculture, and unfortunately, beekeepers can no longer claim to be outside the pesticide system. The common use of Apistan in our colonies has removed beekeeping from the small group of agricultural pursuits that are pesticide-free. We are now active participants in the pesticide problem, and it is a problem. The statistics concerning pesticide impacts on humans are frightening, and indicate that our dependence on chemicals has side effects severe enough to bring into question the cost-effectiveness of using pesticides at all.

David Pimentel and his colleagues at Cornell University have estimated that the detrimental side effects of pesticide use in the United States cost $8 billion per year. This amount includes public health costs, loss of domestic animals and their products, fish losses, monitoring and cleanup expenses, and the cost of enforcing government regulations. One of the largest costs, about $790 million dollars annually, was to treat the estimated ten thousand pesticide-related cancer cases that appear in the United States each year. Worldwide, acute pesticide exposure causes twenty thousand deaths each year, and of course it is almost impossible to put a dollar value on that figure.

The $8 billion pesticides cost the United States in side effects is close to the $8.5 billion spent to purchase pesticides each year in that country. Together, these two "costs" equal the $16 billion in crops that pesticides are estimated to save every year. One final statistic: in spite of using a billion pounds of pesticides annually, we lose 37 percent of our crops to pests, and this value is increasing rather than decreasing.

There is another side to this story, however. If beekeepers suddenly stopped using Apistan, virtually every one of us would be out of business. We are in the ironic position of having to put pesticides into colonies in order to manage bees successfully. I say ironic because traditionally beekeepers have been among the most ardent opponents of pesticide use. The reputation of honey as a natural "organic" product is not consistent with putting chemicals into hives. Unless you're one of the lucky few who have no *Varroa*, however, you have no choice; if you don't use Apistan—or perhaps formic acid if you're a Canadian beekeeper—your colonies will almost certainly die.

This dilemma is not unique to beekeeping. Every farmer has to decide whether or not to use pesticides. Growers are faced daily with the decision to use chemicals or risk their crops. Very few growers, and I'm sure very few beekeepers, are willing to take that risk, and so most of us head to our local chemical supply store and spray, spray, spray at the first sight of pests. This reaction is shared by city dwellers, too, who dash to their local hardware store after seeing a lowly cockroach scuttle across the kitchen floor and buy something, anything, to annihilate the pest. Farmers, at least, can say that they're trying to make a living!

Most beekeepers, and most farmers for that matter, are not horrible spray jockeys poisoning the environment in order to make a few bucks. We use chemicals because we feel there are no other options, and in the case of *Varroa* today, there really *are* none. Given the dependence on chemicals that has rapidly descended on the beekeeping industry, it is important for all of us to keep chemical use in perspective and do all we can to use the least amount possible.

The first and most obvious way to reduce chemical use is to use Apistan only when sampling determines that it is needed. Too many beekeepers follow the practice that developed among an earlier generation of farmers and use pesticides on a schedule rather than only when sampling reveals the need. Today's farmers use monitoring systems to follow pest populations and spray only when the pest is reaching economic thresholds, and then only as much and as frequently as necessary to reduce the pest presence to a noneconomic level.

Beekeepers should do the same with *Varroa*. Some apiaries may need treatment only once a year, others twice, and some perhaps three times. There are numerous quick and easy sampling techniques that beekeepers can use themselves, with no scientific training necessary. A responsible pesticide user samples the bees and makes an informed decision before treating with a pesticide.

Unfortunately, we still don't have a good sense of what mite levels justify Apistan applications. A number of rough guides have been released by extension workers that relate the number of mites found by various sampling methods and advise when to treat based on these samples. These are excellent first steps, but we need more refined information to relate mite levels to economic damage thresholds in order to make better decisions about when to treat.

Safe use of Apistan is highly recommended for your own health as well as to protect the environment. It's not a bad idea to take a pesticide applicator's course, even though it may not be required by your state or provincial officials. Simple safety rules are not hard to follow. For example, always wear gloves when handling pesticides, and a respirator if working indoors. Take the used strips to a proper disposal center when you remove them from colonies, and *don't* use them again. Finally, never store Apistan where children might have access to it.

At one level, the obvious applies: if you need to use chemicals, use them sparingly and safely. Most farmers in North America do use pesticides sparingly and safely, but we're still spraying more than a billion pounds a year of active pesticide ingredients into our environment. Beekeepers, like farmers, need to begin thinking about ways of getting off this pesticide treadmill, and for *Varroa* it won't be easy.

Finding alternatives to toxic pesticides for mite control may be the single most valuable contribution research can make to beekeeping. First, I would look at alternative chemicals that are not toxic to anything but mites. Some researchers are now examining essential oils, substances similar to the menthol now used against tracheal mites, as mite controls, and we should expand such studies. These oils may not work as effectively as Apistan, but they may keep *Varroa* below economic thresholds, which is the control level we should be looking for rather than total eradication.

Perhaps we also should consider more labor-intensive physical controls. For example, breaks in the brood cycle will reduce *Varroa* levels. And drone comb can be used as a "trap" to attract mites and then can be removed and destroyed. These are expensive methods because they involve considerable labor, loss of equipment, and some reduction in colony strength when brood rearing is reduced, but human society as a whole needs to decide how committed we are to reducing pesticide use. If beekeepers are truly against pesticides, perhaps we need to look at more expensive but environmentally cleaner methods. At the least, we can conduct some research comparing the economics of physical and chemical control techniques to determine whether nonchemical methods are viable.

In the end, it comes down to money. If beekeepers paid a twenty-five cent tax on every strip of Apistan used, and devoted that money toward research, I am confident that alternative control methods for *Varroa* could be developed that would either use non-toxic chemicals or avoid chemical use entirely. I estimate that such a fund would generate $500,000 annually—enough to fund numerous tests to screen alternative chemicals and examine the biological efficacy and economics of numerous control techniques. This "tax" would disappear after two or three years, because it wouldn't take very long to solve the mite problem if sufficient funds were available, especially if a panel of beekeepers and scientists was overseeing the work.

I suggest this idea with no expectation that it will happen, because there doesn't seem to be any way to collect such a tax, and most beekeepers would grouse about yet another bill. Pesticide companies certainly would not cooperate in overseeing a "tax" collection that would put them out of business, and beekeepers are too independent-minded to arrange something like this on our own. Nonetheless, think about it for a minute: if each of us gave a quarter for every Apistan strip we used, and beekeepers decided which projects would get the money, within three years we might have methods to control *Varroa* that would not involve putting toxins into our hives. Bookeepers could then withdraw from the "billion pounder" pesticide club. How about it, presidents of beekeeping organizations: why not take up the challenge of making this work? Maybe, just maybe, it would be worth it.

12. Semiochemicals and Varroa

THE MOST SIGNIFICANT ISSUE IN BEE RESEARCH today is a seemingly simple problem: How can we control the *Varroa* mite? Fortunately, there is at least one solution to the problem, Apistan. Although expensive and tedious to apply, it does kill most *Varroa* mites without harming bees or leaving undue residues, if properly applied. Apistan, however, has some real and potential problems associated with it. The real problem is that it is a pesticide, and none of us likes putting chemicals into our hives. The potential problem is resistance; mites are notorious for developing resistance to pesticides, and resistance to Apistan has already been reported in Europe. Thus, new approaches to *Varroa* monitoring and control should be a primary focus of the bee research community, particularly approaches that do not involve hard pesticides.

One such method may involve semiochemicals, and is a field of research that a number of laboratories may enter in the near future if funds become available. The term *semiochemical*, from the Greek *semion*, to sign or signal, was coined to reflect the broad scope of modern scientific inquiries into chemical communication. The existence of semiochemicals and their potential for managing insects and other pests have been known for some time. The French entomologist J. H. Fabre was the first to formally investigate insects' ability to use chemicals to find each other across long distances. In classic experiments, he put female moths in wire cages, placed these on his windowsill, and then observed that tens or even hundreds of males were attracted to the cages. Even more remarkably, when Fabre and his colleagues marked male moths and released them at distances up to eleven kilometers away from the caged females, many of the released males appeared at the cages within hours. When their antennae were removed or painted over with lacquer, however, the males lost their ability to find the cages holding the females, even over short distances. In a book titled *Bilder aus der Insektenwelt* Fabre speculated that male insects, especially moths, orient to scents released by females, and predicted that one day "science, instructed by the insect, would give us a radiograph sensitive to odors, and this artificial nose will open up a new world of marvels."

The first isolation, chemical identification, and synthesis of a pheromone did not occur until the late 1950s, when gas chromatography was perfected by chemists. This technique separates compounds and allows them to move through a column. Each compound moves at a distinctive rate. German scientists used this new technique to elucidate the sex pheromone produced by the female silk moth *Bombyx mori*. The silk moth was an unlikely candidate to initiate the field of insect chemical ecology. It is not a pest but a beneficial insect, and there was no compelling economic reason to find the female's sex attractant. This moth had one advantage over the pest species that might have been chosen for the honor of being the first insect to divulge the identity of its aphrodisiac chemicals, however: it is large, and in the 1950s the technology to identify minute quantities of insect-produced chemicals was in an early and crude state.

Even so, the task of isolating enough of the chemical to identify was daunting. The mating ritual of the male and female moths is well known, and it provided a good bioassay for potential pheromonal compounds. The female sits on a tree trunk, everting a gland in her abdomen and releasing the attractant pheromone. The male flies upwind, using his large, plumose antennae to detect the female's species-specific scent and orient to her. Subsequent studies showed that the male antennae respond to as little as one molecule of attractant, and can locate a female even if only a few hundred molecules are contained in her odor plume.

Unfortunately for science, each female produces only about one millionth of a gram from her abdominal gland, enough potentially to attract a billion males, but far below the detection capabilities of 1950s technology. The Munich scientists, led by A. J. Butenandt, had to clip 500,000 female abdomens to extract enough of the attractant to identify its chemical structure, but they finally succeeded in 1959. They named the attractant odor bombykol, and found that when a synthetic version of bombykol was placed on a lure, it attracted males as well as a live female moth did.

The potential impact of identifying the silkworm sex attractant was not lost on the scientific and pest management communities. It was immediately obvious that pheromones could be used as management tools to overcome insect pests with their own compounds. The 1960s saw a trickle of new chemicals isolated, identified, synthesized, and then tested. The trickle grew to a torrent as techniques improved, instruments became more sensitive, and growing knowledge of semiochemical-based biology created an increasingly sophisticated substrate for subsequent researchers to build on. The growing interest in semiochemicals, and our increasing technical capability to identify them, is reflected in the number of U.S. patents granted for novel compounds. Only 13 patents were granted before 1970, but 150 had been granted by 1988, and 257 by 1991.

The interest in mating attractants gradually expanded into the discipline of chemical ecology, which includes not only mating substances but any chemical involved in communication between organisms. Today's chemical ecologist might still elucidate the identity and function of an insect sex attractant but is just as likely to study the odors that attract a pine beetle to its host tree, the inhibitory secretions that prevent a worker bee from laying eggs, or the alarm chemicals given off by an aphid under attack by parasitic wasps.

The work of these chemical explorers has been of considerable interest to pest managers, because semiochemical-based pest management has great potential advantages over more traditional pesticide-based control. Most significant, semiochemicals are highly specific to individual species, active at sometimes unbelievably low concentrations, and are relatively easy to register and market because they have virtually no side effects on vertebrates, or even on other insects.

The practical uses of semiochemicals in pest management have settled out into three main areas: monitoring, attracting and killing, and mating disruption. The first two have excellent potential for use in *Varroa* management. If we could find an attractant for *Varroa* mites, it would be possible to monitor their presence and levels in hives, and perhaps even attract most of them to a trap in which they could be killed.

The first step in a semiochemical-based control program is to find compounds attractive to *Varroa* mites. We already know the mites are attracted to a number of identified compounds that honey bee larvae release just before capping. These compounds signal adult workers to construct wax caps over the larvae's cells. Since *Varroa* mites enter the cells just before the capping, these compounds would be ideal attractants in a monitoring or control program. Further, *Varroa* mites are preferentially attracted to drone larvae and to adult nurse bees, providing additional possibilities to isolate, identify, and synthesize *Varroa* attractants.

Once we have an attractant, the next step is to develop monitoring and control traps. Monitoring is the most common application of pheromones in pest management. Typically, an open trap with a lure inside baited with the target mite or insect attractant is set out. The *Varroa* mite would enter the trap expecting to find a larva or adult bee, but instead would encounter a sticky lining from which it could not escape. Beekeepers could check the traps on a regular basis, correlate the numbers of trapped insects with potential economic damage, and make informed decisions about when and how often to apply miticide treatments.

Even better would be a trap into which most of the *Varroa* mites infesting a colony could be attracted and killed. The mites would enter traps baited with attractant and meet a deadly contact poison. This technique has been particularly successful against insects in enclosed spaces, such as beetles in grain bins and cockroaches in interior urban settings. The advantage of this technique over the current method using Apistan is that the miticides could be contained within traps and would not come into contact with the bees, comb, or honey. Even better would be a trap that worked via a sticky lining alone, without any miticide being needed.

Is semiochemical-based control of *Varroa* really feasible? Semiochemicals are becoming more popular and effective as scientists and pest managers become more familiar with the techniques that use them, and there is no reason to believe that managing *Varroa* with semiochemicals would be more difficult than managing other pests. The semiochemical approach certainly seems to have potential, even if it doesn't ever go beyond providing an inexpensive and easy-to-use technique for monitoring mite levels. The only way to find out whether the potential can be realized is to try it. That, after all, is what research means: "the act of searching carefully for a specified thing." My own feeling is that a careful search for an attractant-based monitoring and control system for *Varroa* will be successful. Beekeepers and chemical companies might consider sending their research dollars in that direction. After all, *Varroa* is our most important problem, and semiochemicals offer the best alternative to putting hard chemical pesticides into our colonies.

13. Killer Bee Killers

THE GRASS IS NOT ALWAYS GREENER ON THE other side. In this case, I'm thinking of the other side of the world, South Africa, which has a beekeeping problem far worse than any problem beekeepers face in North America today, and that includes mites, Chinese honey, and Africanized bees. South African beekeepers not only have the original version of the "African killer bee," they have another killer bee that kills the killer bee! I'm referring to the Cape bee (*Apis mellifera capensis*), a race of honey bee that has been devastating South African beekeeping. The honey bee used in most of South Africa is the African bee (*Apis mellifera scutellata*), the same bee that was imported into Brazil and has now arrived in the southern United States. This bee is difficult enough to manage, but *capensis* has compounded that difficulty exponentially.

The Cape bee was once confined to the region at the tip of South Africa and showed no inclination to migrate north on its own. However, beekeepers in the early 1990s began moving the African bee south to the Cape for crop pollination, and then back north for the rest of the season. Evidently, quite a few Cape bees got into the African colonies, were moved north, and have now become resident there. This is a major problem for South African beekeepers because Cape bee workers can lay eggs that develop into females, even though the workers don't mate. That by itself might not be so bad, but these laying workers will take over a colony and eventually kill the colony's queen, and then the colony dwindles and dies. Sometimes eggs laid by these laying workers are reared into queens, but then the colony swarms or absconds. Either way, the beekeeper is left with a dwindling and eventually dead unit.

The worst aspect of this problem is that there does not seem to be a solution. I recently participated in an electronic mail discussion group on the Internet that was started by the South Africans to solicit input from researchers around the world. The best minds in bee science drove this problem around the electronic highway for a month. Lots of questions were asked, but no solutions emerged. This is a fascinating research area, but the research has been going on for some time and has not yet helped the beekeepers. A large proportion of South Africa's bee colonies are affected each year, and the term *devastating* is not out of proportion to the impact the Cape bee has had in the region.

Not surprisingly, there was considerable discussion among the Internet group concerning how the Cape bee was moved out of its original, and very narrow, range, and why it succeeded in expanding northward when it had not done so before. The *how* seems fairly well established. The fruit industry in the south Cape region expanded. Eventually the local bee colonies were not sufficient to pollinate the developing crop, and pollination-oriented migratory beekeepers were hired from the more extensive beekeeping regions in the north. Thousands of commercial African bee hives were brought down to pollinate the fruit trees, and Cape bee workers entered some of those colonies before they were moved back north. The problem was compounded because many South African beekeepers don't requeen with commercially reared queens, but instead split colonies and let them requeen themselves. As Paul Magnuson from South Africa put it in one of his e-mail messages, "This was a recipe for disaster. African colonies that had contact with the Cape bee began to develop strange symptoms: mysterious queen loss, laying worker activity, decrease in foraging, lack of the normal defensiveness, and general demoralization."

These colonies did indeed requeen themselves, but with laying workers rather than with mated queens. A laying Cape worker is different from beekeepers' usual concept of a laying worker because she can lay female eggs, even without mating. A typical Cape bee takeover of an African colony proceeds as follows: Some Cape workers drift into the African colony, fight with the resident African bees, and lay eggs. The Cape bees kill the African queen and continue to lay eggs for many months. Eventually, the colony becomes completely Cape, dwindles, and either dies or rears a Cape queen. The Cape queen route doesn't solve the problem, however, because even queenright Cape bee colonies are not very productive. Rather, they continue to requeen themselves every few months, and rarely reach a very large population.

This is not a trivial problem. About 150,000 colonies have been killed either directly as a result of Cape bee infestation or through legislation that requires "depopulation" of Cape bee colonies outside their natural range. Most of the honey in South Africa is now imported. South African beekeepers are committed to finding a way to return to beekeeping with the African rather than the Cape race of bee, but this will not be as easy as it might seem. The Cape bee is so widespread now throughout South Africa that nothing short of killing virtually all the commercial colonies in the country has any potential to clean up the mess. Even then, the Cape bee may now be endemic in feral colonies and could easily repopulate a new batch of commercial bees.

The Cape bee problem has generated interesting research opportunities and is an area in which bee research could have an enormous beneficial impact on the bee industry. Perhaps the most interesting question is not why is the Cape bee causing problems, but rather why didn't it cause problems before? There are no dramatic geographic or climatic barriers to the spread of the Cape bee, yet only recently have these bees become established outside the very tip of Africa. The standard explanation has been that the two races are ecologically adapted for their respective regions and cannot expand beyond them. This theory was tenable before the Cape bee began doing well in the African bee areas, but the Cape bee's success in the north poked a huge hole in it.

Another fascinating research topic concerns why only Cape bee workers commonly lay female eggs. Laying workers of other bee races occasionally lay female eggs, perhaps one egg in a thousand, but a Cape laying worker does it all the time, and in fact does not lay many drone eggs. Is this a primitive trait that has almost disappeared in all other honey bee races or a recent development in bee evolution that soon will spread throughout the world? I suspect it's a primitive trait that survives only in this small Cape population, because it reflects a fairly uncooperative social situation in which workers and queens are fighting for the right to reproduce. In the rest of the honey bee world the queens and workers are further along the social spectrum toward cooperation, which seems to me a more advanced trait.

Another good question is why this trait persists at all in the Cape bee, regardless of whether it's advanced or primitive. According to one explanation, the southern Cape is very windy and queens are frequently lost during mating flights, and a colony's ability to requeen itself through laying worker eggs might be very strongly selected. Honey bees exist and mate in other parts of the world with high winds, however, yet this trait has not become established anywhere else. No, the reasons behind the evolution of workers that can lay female eggs and the isolation of this characteristic in the southern Cape region until recently remain a mystery.

With regard to what to do about the Cape bee, the world's bee brains, myself included, seemed at a loss as to what to recommend. Lots of ideas flowed back on forth on the Internet, but we really don't seem to know enough about the Cape bee to find the magic bullet that will return South African beekeeping to a healthy and profitable state. Each of us applied our own expertise to the problem. Geneticists suggested research on how laying worker eggs develop into females, population biologists proposed studies to better understand hybridization between the races, and pheromone biologists such as myself suggested that the answer to the Cape problem might lie in a better understanding of chemical communication between workers and queens.

All of us are probably right. No single answer is going to solve the problem, but a broad understanding of this interesting bee's biology may lead to a gradual improvement in South African beekeeping. If I were still a young student, I think I would head to South Africa and begin working on the Cape bee. The research opportunities there are among the most fascinating on the planet and have great potential to contribute to the survival and prosperity of South African beekeeping. If any one thing came out of the discussion group as far as research is concerned, it was that there are still research problems in bee science that are both biologically interesting and economically important. And if any one message emerged for North American beekeepers, it was this: DO NOT, under any circumstances, import bees from South Africa. If we didn't learn anything from African bees being imported into the New World, we would certainly learn from the Cape bee that the grass is not greener somewhere else.

14. Bee Nutrition: A Dead Science?

LET ME HEAD OFF THE WAVE OF "YOU'RE WRONG, you idiot, what do you professors know about beekeeping?" letters that this essay could inspire by stating right at the start, categorically, unequivocally, and without a doubt: I believe in pollen supplement feeding. I feed my bees with a pollen supplement every spring, and sometimes at other times during the season. I consider supplement feeding one of the most important aspects of bee management, so do not interpret this essay to mean that you should not feed pollen supplements. The commercially available supplements do a reasonable job of stimulating bees and increasing colony populations.

That said, I must also say I think we can do better. Bee nutrition seems to be a dying or even dead art; the last wave of good research on bee nutrition ended in 1985. There are virtually no research scientists today working to refine our knowledge of the necessary components of bee diets and improve the available supplements. Although the supplements we now use work, I think there is considerable room for improvement in what we feed our bees.

There have been two "great" bee nutritionists in our time, Mykola Haydak and Elton Herbert. The early and most classic work, elegantly performed by Haydak, was published in a deluge of papers that extended over a period of five decades, from the 1930s into the 1970s. Haydak's work literally defined the field. He rigorously examined the nutritional requirements of bees, particularly of brood, and extended that work into the development of pollen supplements for beekeeping applications. His intellectual successor, Herbert, picked up the torch in the 1970s, and further defined bee nutritional requirements and investigated supplemental feeding. Many others have worked in this field as well, of course, but these two distinguished scientists have been the most significant contributors. Unfortunately, the untimely passing of Herbert in the mid-1980s seems to have brought the field of bee nutrition to a grinding halt.

Bees require two major components in their diet: sugar and protein. Sugars are needed primarily for energy and come mostly from nectar. Proteins are the building blocks bees use to construct body tissue, glands, blood, and so on. Without protein—and in fact without the right types of proteins—larval bees cannot grow and adult bees cannot perform many of their normal tasks. The larvae require a considerable amount of protein to grow and develop into adults, and adults require protein to complete their development following emergence, produce brood food and wax, and repair worn-out muscles and internal tissues. The only source of protein naturally available to bees is pollen, which normally contains anywhere from 6 to 28 percent protein.

Pollen also contains most of the other components necessary for a good bee diet, including lipids, vitamins, and minerals. Although lipids are not generally considered major nutrients for insects, some of the lipids in honey bee brood food are essential. For example, bees need cholesterol in their diet but cannot make it themselves, so they must obtain it from pollen. Although cholesterol and related sterols make up only 0.25 percent of brood food, bees will not develop properly as larvae or function properly as adults without it. Some of the B, C, and other vitamin complexes are required for proper growth and development, and again, bees deprived of these materials will not develop properly. Regarding minerals almost nothing is known, although the addition of potassium, sodium, calcium, and other minerals in the form of pollen ash seems to improve the adult bees' ability to rear brood.

Once the basics of bee nutrition were understood, the information was used to design protein-based supplemental feeds for bee colonies. Numerous human foods and food by-products can be eaten by bees when properly formulated, and a number of successful commercial formulations are based on substances such as brewer's yeast, Torula yeast (a type of brewer's yeast), expeller-processed soya bean flour, fish meals, skim milk powders, and Wheast (a by-product of cottage cheese production). These materials generally contain more than 50 percent protein, but need special processing to be suitable for bees. Some need to have fats and/or salt removed, and dairy products can be poisonous for bees unless the milk sugars lactose and galactose are removed. Thus, it is important to use only those products whose labels indicate that they have been properly formulated for bees.

Pollen supplements are usually fed to bees in moist patties composed of the supplemental feed mixed with thick sugar syrup to form a dough, often with a small amount of pollen added as an attractant and feeding stimulant. The patties are placed on the top bars of the brood chamber and covered to prevent them from drying out. Supplemental feeding is most commonly used to stimulate colony growth early in the season, before fresh-collected pollen is available, and is particularly useful when premature spring colony growth is required for package bee production, spring pollination, or early spring honeyflows. Pollen supplements are effective because they stimulate nurse bees to begin producing brood food and provide surplus protein that these nurse bees can use to rear additional brood. The end effect of supplemental feeding is to increase the colony population earlier than would naturally occur.

Surprisingly little is known about the economic value of pollen supplement feeding, but the few studies that have been done suggest that it can be an effective management tool. For example, in 1976 Herbert and his colleagues demonstrated a twenty-five-pound increase in honey production in packages established on March 1 and fed a Wheast-based supplement over packages not fed a pollen supplement. Economically, the Wheast-fed packages earned about twenty-five dollars more per colony in 1998 dollars when feed costs were deducted, although the researchers did not consider the cost of labor. Interestingly, similar supplement-fed packages established on April 15 produced significantly *less* honey than control packages, indicating the importance of proper timing in supplement feeding.

Research conducted by Keith Doull in Australia in 1980 showed that supplement feeding increased honey production by close to 40 percent, although Doull used only five colonies per treatment. A study done in my own laboratory demonstrated that feeding approximately one dollar's worth of pollen supplement in February resulted in increased bee populations relative to unfed colonies and a ten-dollar profit increase for package bee producers.

These apparent successes of supplemental feeding are excellent examples of how basic research can be exploited for beekeeping applications, yet there are a number of troublesome aspects to supplemental feeds that seem to call for additional study. The first thing to consider is that we have not been able to develop a pollen substitute, only pollen supplements. Pollen is still the only complete bee food, and no supplement can replace fresh pollen for more than a few weeks before colonies begin to show nutritional deficiencies.

I suspect we're doing a reasonable job of replacing the abundant proteins of pollen but are missing essential vitamins and minerals that may be present in minute quantities. In fact, we know very little about the role of vitamins and minerals in bee nutrition. There have been numerous advances in food technology and in our understanding of the roles vitamins and minerals play in the nutrition of other organisms, and studies using honey bees would be a fertile and useful avenue of research for a student looking for a place in honey bee science.

I also question whether we have adequately examined the effects of supplemental feeding on individual bees, or the economics of feeding supplements in different situations. Perhaps we have been lulled into a false sense of success by the obvious increase in bee populations induced by supplemental feedings. While we may be increasing bee numbers, these population increases could be occurring at the expense of high-quality bees, and may not always be justified.

There is some evidence that supplemental feeds produce bees of lower quality than bees fed only fresh pollen. For example, in a study whose objective was to compare brood survival and adult longevity in colonies fed various supplements, we found that brood survival was equal or better in supplement-fed colonies compared with control colonies that received only fresh bee-collected pollen. However, the life span of adult workers fed supplement as larvae was about four days shorter than that of workers that received only fresh pollen. Unfortunately, we did not carry these studies further to examine colony performance, individual worker behaviors, or the relative economic impact of the supplemental feedings. Nevertheless, our results suggest that pollen supplements may increase worker populations, but with workers of lower quality. This aspect of supplemental feeding needs more study.

There are other reasons to suspect that even the best pollen supplements produce individual workers of diminished quality. Subtle deficiencies in types and amounts of proteins and lipids are known to induce lower-weight bees, inhibit proper larval development, and interfere with the completion of adult development, particularly adult glandular development. We assume that such deficiencies are compensated for by larger bee populations, but this trade-off needs further study.

Another useful project would be to identify the natural substances found in pollen that attract bees to supplements in the hive and stimulate feeding. Anyone who has made up supplements without adding some pollen to them may well have found dried-out, unused patties in the hive a few weeks later because the bees were not attracted to the feed. However, we don't always have pollen available to add to patties, and purchased pollen may transmit disease. Thus, synthetic attractants derived from the natural attractants in pollen would be very useful in supplement formulations.

I have tried to be both positive about the benefits of supplement feeding and cautious about assuming we know everything there is to know on the subject. I believe we can formulate more attractive and nutritious supplements that not only induce larger colony populations but also provide the same high-quality workers that result from brood fed fresh pollen. It's time to revive the art and science of bee nutrition, and couple it with economic management studies to provide beekeepers with improved formulations of what can be a very potent management tool, pollen supplements.

15. Tracheal Mite Research: The Next Generation

REMEMBER THE HONEY BEE TRACHEAL MITE? THE discovery of this pest in the United States in 1985 clearly affected both honey bees and the politics and economics of how we manage them. Study after study showed heavy winter losses due to tracheal mites, which led to intensive and expensive testing for the presence of these parasites in colonies, some quarantines and restrictions on bee movement, and severe blows to the American package bee and queen industries when Canada placed certain restrictions on the importation of bees from the United States. Eventually, treatments such as menthol, formic acid, and various vegetable oil patties were developed to combat the tracheal mites, and *Varroa* and Africanized bees appeared. Today, tracheal mites have been relegated to relatively minor notoriety, if not in their impact, then certainly in the attention they are receiving from beekeepers and researchers.

Curiously, although we can now control the tracheal mite chemically, we still have very little understanding of its biology and impact. No one has yet discovered how tracheal mites affect bees or why they appear to cause serious economic damage in some colonies and locations but not in others. We have chemicals that we apply for treatment, yet we don't know how or why they work. There has been considerable research into genetic resistance to tracheal mites, but the mechanism of such resistance in honey bees is unknown, and the economic impact of so-called resistant stock has yet to be proven.

Perhaps the most serious gap in our understanding of tracheal mites is our failure to learn what they do to the individual bee. One theory is an obvious physical one: a bee carrying a heavy mite infestation in its breathing tubes should have trouble breathing, leading to poor flight performance and a short life. This is not the case, however; infested and uninfested bees show no difference in performance characteristics such as the number and timing of foraging trips, nectar load size, and length of life. Another theory is that tracheal mites themselves are fairly benign, but they transmit some bacterial or viral disease that weakens and kills bees. This is an area of active research, but no clear connection has been established between tracheal mites and bee diseases.

Our failure to discover just how tracheal mites damage honey bee colonies is particularly remarkable because we have developed numerous control methods that seem to work. This is unusual in pest control. Generally, management tools are developed to target particular aspects of a pest's behavior or physiology. We dump aromatic chemicals (menthol), caustic compounds (formic acid), and benign substances (vegetable oils) into colonies for tracheal mite control, and they reduce the number and impact of mites, at least when properly applied, yet, we have little idea why these substances are effective! We have bypassed the typical research protocols in pest management, which almost always begin by first understanding a pest's biology and mode of action, and then developing a control measure that affects the pest without hurting the host animal or plant. It would be nice to say that the bee research community has been extraordinarily astute in determining this pest's weak points and developing control measures, but in fact I think we have been exceptionally lucky to find control methods in the absence of any significant knowledge of this pest's mode of action.

The real impact of tracheal mites is another subject of confusion in the beekeeping community. Most studies, particularly those in northern regions of North America, have shown high levels of winter loss and poor colony performance in tracheal mite–infested colonies. Indeed, infestation rates of 10 percent or higher are considered serious danger signs, and virtually all extension agencies in North America recommend annual or more frequent treatments when mites are present. Yet, we continue to hear reports from beekeepers who claim that although they have tracheal mites, the mites are not an economic problem; and most authorities in Europe claim that the mites are not a significant pest. Indeed, European recommendations for bee management generally do not include measures for tracheal mite control.

One explanation for these divergent reports is that there is regional variation in the virulence of the mites. Of course, we cannot make much progress in analyzing this possibility until we understand what mites do to bees, but such variation in virulence is typical of pests and diseases. The best strategy for a parasite such as the tracheal mite is not to kill its host, but rather to live within the host while causing only minimal damage. When the host dies, so does the parasite. Clearly it would be advantageous for tracheal mites to inflict only minor damage on the host bees. This explanation, although pure speculation, makes sense and deserves some attention from the research community.

The more common explanation for the reports of varying tracheal mite impacts on bees is that some honey bee colonies are resistant to tracheal mites. The possibility of genetic resistance has led to numerous selection programs designed to find and breed mite-resistant bees. These programs are quite expensive to conduct, and frequently involve importing foreign bee stock (under quarantine, of course) on the assumption that European bees have some resistance mechanism that is not present in North America.

I have been somewhat critical of our approach to finding and breeding tracheal mite–resistant stock because I think we have been going about it in the wrong way. First, we have been focusing on finding "resistance" but paying little or no attention to the bee characteristics that might impart it, or to what that resistance might mean at economic levels.

Another problem with resistance research is that the standard selection protocols have not been taken to the final step, production-level studies. The procedure for determining resistance was originally developed by Norm Gary, Rob Page, and their colleagues at the University of California at Davis and Ohio State University. Their protocol involves placing groups of test bees into the same colonies or cages, and then counting the number of mites per bee from each group a week or two later. Presumably, the lines of bees with fewer mites are more "resistant." This approach is fine as a first step, but we have not yet taken it to the really important level: performance of colonies of so-called resistant stock. A Canada-wide study in the early 1990s demonstrated that colonies selected for mite resistance did indeed have fewer mites than unselected stock, but the selected and unselected colonies were no different in other characteristics important for commercial beekeepers. Resistant bees may carry fewer mites under highly controlled research conditions, but may not produce any more honey or survive cold winters any better than bees selected for commercial attributes other than disease resistance.

Nor do we know what resistance might mean in terms of chemical treatments. Can so-called resistant stock be managed without chemicals, or is some level of chemical control still required? If chemical treatments are required, is the resistant stock productive enough to justify its use? These are all questions that should be the bottom line of resistance research, but have not yet been addressed.

Finally, my particular pet peeve concerns the expensive and laborious importation of potentially resistant stock under complex quarantine conditions. The rationale for such importations is that European bee stock supposedly is not affected by tracheal mites, so North American beekeepers should incorporate European genetic material into our North American stock. I have argued against such importations for a number of reasons. First, North American stock that tests equal to or better than any stock that has been imported, at least using the mite load tests described above, is available. Thus, the elaborate and time-consuming protocols involved with importation may simply not be necessary.

Second, and perhaps more important, there is no reason to believe that even resistant European stock will perform well under North American conditions. Beekeeping in Europe is very different from beekeeping in North America. There is relatively little large-scale commercial beekeeping there, almost no migratory beekeeping, and certainly the climatic and floral conditions are different. Although our North American stock originated largely in Europe, it has been intensively bred for our beekeeping conditions, and likely differs substantially from the original stock in its management characteristics. I think a better selection strategy would be to focus first on identifying good North American beekeeping stock, and then examining these bees for resistance.

I also question whether research-oriented breeding programs have much chance of contributing to long-term stock availability. Selection and breeding is a long, complex, and expensive process, and is best conducted by government and university researchers for those reasons. However, our research funding structure today does not encourage stock maintenance after the selection studies are completed. Even though U.S. and Canadian governments have sponsored many selection programs, almost none of the breeding stock the programs developed has been maintained for any length of time or made its way into widespread commercial use. This is not to fault the research community, or to downplay the potential of such selected stock. Rather, we always seem to put the cart before the horse in selection programs. That is, selection receives funding and research attention, but once selected stock is available, no provisions are made for long-term maintenance and distribution of it.

Where does this leave us with tracheal mites? Researchers need to focus on how mites affect bees in order to develop better management tools, including, possibly, natural resistance. And our resistance studies need to focus more on North American stock that has proven useful for beekeeping than on European stock of unknown merit. Finally, resistance development will be only of academic interest unless methods of stock maintenance and distribution are developed, and we need to determine if and how this will be accomplished *before* investing more research funds into selecting for mite-resistant stock.

16. Mite Load

THE SPRING OF 1996 WAS A REAL GOOD NEWS/ bad news season for beekeepers. The good news was that honey prices reached record levels and appeared to be staying high. The bad news was that beekeepers in many states didn't have enough colonies to make money on the windfall honey prices. Many beekeepers reported 50 to 80 percent colony losses over the previous winter, with beekeepers in the northern United States the most affected. The standard explanation for this high winter loss was parasitic mites, specifically the combination of tracheal and *Varroa* mites.

I'm not so sure. That is, I believe the mites played a major role in the winter loss, but I'm not sure they deserve all the blame. Both tracheal and *Varroa* mites can be kept at levels low enough to have only minimal effects on colony health. We have the medications and techniques available to deal with the mites, and proper applications should be effective. Yet, beekeepers throughout the United States have been reporting major losses. What's going on?

I think the high winter loss is the result of a mix of colony management problems that we are only beginning to appreciate, inadequate extension information for beekeepers on how to deal with diverse mite situations, and a substandard inspection and regulatory process that has left U.S. beekeepers with insufficient tools to cope with parasitic mites. This combination of factors is a potent brew that has left beekeepers struggling with a problem they shouldn't have.

One problem with tracheal and *Varroa* mites is that the two mites occurring together may be considerably more problematic than either is separately, especially if colonies are stressed in other ways as well. My research group experienced this firsthand a few years ago, when beekeepers in our region lost 66 percent of their colonies one winter, and we lost about 30 percent of our university bees. At that time, tracheal mites had been around for a few years, but were present in low numbers and seemed to be a minor problem. The previous summer had been rainy and cool, and *Varroa* arrived, although the *Varroa* mites were not very abundant going into the winter. The combination of colonies at below-average strength afflicted with low levels of both mites proved deadly, however, and winter losses were high.

This experience illustrates the concept of synergy, a situation in which two or three factors taken separately have little effect, but when combined become much more serious than simple addition of their effects would predict. Thus, strong colonies going into winter with low populations of one mite could be fine, and below-average-strength colonies with no mites might still do reasonably well, but slightly weakened colonies with both mites present at low to moderate levels may be a disaster.

Two of my students are currently investigating this phenomenon. Danielle Downey is comparing colonies with no mites, tracheal mites only, *Varroa* mites only, and both mites together, examining individual bee behavior and colony-level effects. Her early findings seem to confirm that the bees in colonies with either mite are less effective foragers than those from mite-free colonies, especially for pollen, but this project is still in its early stages. Alida Janmaat has been comparing the effects of *Varroa* mites on well-fed colonies and colonies deprived of some pollen. Again, her results are just coming in, but they suggest that colonies weakened by poor nutrition suffer more from *Varroa* infestations than well-fed colonies do. If these results are correct, then simple mite control may not be sufficient to guarantee colony survival. Rather, good overall colony management that includes the maintenance of populous colonies going into the winter, control of other diseases, and an adequate level of nutrition is a critical factor in determining beekeeping success.

Another problem that may afflict many colonies is improper application of mite control measures by the beekeeper. Whatever the method used—be it grease extender patties, menthol, Apistan, or formic acid—these tools are expensive and labor-intensive to apply. It is not surprising that some beekeepers do only the bare minimum for mite control, but perhaps we have been underestimating the importance of continued vigilance against mites.

A beekeeper running one thousand colonies may be tempted to skip the spring treatment with menthol if the colonies seem strong, or perhaps to pass on the fall treatment with Apistan if *Varroa* mites are not visibly abundant. It is easy to underestimate mite levels unless careful sampling is done, but sampling is expensive, and cutting back on sampling may appear to be an attractive way to cut costs when colonies seem healthy. Unfortunately, it doesn't take more than one cost-cutting reduction in sampling or treatment to allow mite levels to boom, and suddenly a 50-percent spring loss is staring you in the face.

Even vigilant beekeepers are faced with a problem, because the extension information available today does not adequately define when and how to treat for one or both mites. There is an overabundance of sampling methods to choose from, and a poor understanding of the relationship between mite levels found by the various sampling methods and recommendations for treatments. We are just beginning to see extension publications with clear recommendations for treatments based on pretreatment sampling. We need to further refine these recommendations so they are easier to use and more effective at recommending which treatments should be applied under different conditions.

Another factor that may be producing undesirable levels of winter loss is *Varroa* resistance to Apistan. This is still undocumented in North America, but it is disturbing that beekeepers who do appear to be using Apistan properly, and apply it two or even three times a year, are still experiencing high colony loss. Even more disturbing is the epidemic of improper use of Apistan as well as other, non-registered miticides, at least if the corridor talk at beekeeping meetings is to be believed. Even proper use of Apistan is resulting in high residue levels in beeswax, creating problems for the international beeswax market as well as providing the type of continuous, low-level exposure that can quickly induce resistance. If resistance to Apistan is developing, we are in for a rocky ride the next few years, and the only ray of sunshine to brighten the picture is that honey prices will remain high due to shortages induced by high winter losses.

Although illegal use of miticides by beekeepers may be part of the developing resistance story, the U.S. regulatory process must also shoulder some of the responsibility. The best way to ensure that *Varroa* will develop resistance is to have only one treatment licensed for use against it, which is currently the situation in the United States. Alternating treatments is the best way to prevent resistance, and there *is* an alternative to Apistan that has yet to be registered for use on bees in the United States.

That alternate treatment is, of course, formic acid, which has been registered and used successfully in Canada since 1993. Formic acid is not a panacea for mite problems; it is difficult to apply, can be dangerous to the applicator, is labor-intensive to use, and is not quite as effective against *Varroa* as Apistan. However, it does have two significant advantages. First, it is effective against both tracheal and *Varroa* mites. Second, and most significant, it can be alternated with Apistan, thereby minimizing the possibility that resistance will develop. Canadian beekeepers have had success using Apistan in the fall and formic acid in the spring, sometimes with the addition of an extender patty in the spring or fall to further suppress tracheal mite levels.

It is ironic that formic acid is registered for use in Canada but not the United States, because Canada is usually the stricter of the two countries about licensing chemicals, and requires more testing. It is not clear to me why registering formic acid for use again *Varroa* has been so difficult the United States, especially in view of the fact that low levels of formic acid are naturally found in honey, formic acid leaves little or no residue in wax, and many more toxic pesticides are registered and widely available for other purposes. A well-coordinated, cohesive lobbying effort by the major U.S. beekeeping and professional organizations might overcome the apparently unreasonable caution displayed by U.S. regulatory officials concerning the use of formic acid as a miticide in bee hives.

It may also be time to take a look at reestablishing beekeeping regulatory and inspection services, either government-run or private. Regulation and inspection have declined substantially in the last few years, due partly to government cutbacks but also to beekeepers' desire for less government involvement in inspection and regulation.

With colony losses of 50 to 80 percent in many areas, I think beekeepers could use some help. Increased colony inspection, mandatory treatment for problem apiaries, and better oversight of treatments before colonies are moved might be useful in reducing mite impact. The robust inspections we used to have for American foulbrood, coupled with strict regulations concerning treatments of infested colonies, were effective at bringing AFB under control, and a similar service might do the same for our mite problems.

None of us likes regulations. We have become intolerant of inspection services, and certainly beekeepers would object to paying a fee for inspections. Nevertheless, losing 50 to 80 percent of our bee colonies is not going to be very popular, either. A better-coordinated effort to manage mite and other colony health problems might help reduce this high level of colony loss. And it might be cost-effective. Paying five dollars a year per colony for a government-run or private inspection service that would monitor mite levels and prescribe treatments might seem excessive, but saving ten dollars a year by using only necessary and properly timed applications might be one benefit. Even more significant, a beekeeper using a properly run inspection service could earn hundreds of dollars in honey sales from a colony that might otherwise have died. Seems cost-effective to me.

17. Beekeeping and Snake Oil

I RARELY "SURF THE WEB," AND AM STILL OLD-FASH-ioned enough to prefer going to the library to do research instead of connecting to the World Wide Web to do an information search. For me, there is something more substantial about a fact garnered from a dusty book on a library shelf, or more exciting about an article in the latest peer-reviewed, hard-copy scientific journal, than the ephemeral and unfiltered flood of information that clogs the computer highway.

I recently felt vindicated in my biblio-snobbery after reading an essay that came to me over the Internet. It was written by Tom Sanford of Florida for his excellent monthly newsletter—which, yes, is distributed by electronic mail. He was writing about the use of essential oils such as wintergreen for mite control, and made the point that many beekeepers were trying this and other methods without waiting for scientific testing, let alone proper registration or carefully evaluated descriptions of application methods. He viewed this as a revolution in research and information delivery, because beekeepers were testing mite control methods themselves, sending their observations out over the Internet to other beekeepers, and even describing in detail how to use these illegal controls. The normal filter of rigorous scientific analysis and carefully considered extension recommendations is thus bypassed, replaced by a marketplace of ideas with only beekeepers' experience to filter out good ideas from bad, and to determine what works and what does not.

I am concerned about this trend toward increased information flow with no signposts to guide the unwary Web surfer away from flawed information. It is not that I don't trust beekeepers to make their own management decisions. To the contrary, I have great faith in the beekeeping marketplace to determine which management methods work at economically acceptable levels. Rather, I am concerned that beekeepers may get burned by using unproven methods that don't work, that control techniques with excellent potential may be dismissed too easily because the proper application timing and dose have not been determined, and that improperly used chemicals may leave residues in honey and wax that could damage consumer confidence in our products.

My fear of beekeepers getting burned by relying on improperly tested products is a well-founded one, and I believe the high colony wintering losses reported from the 1995–96 winter season may have been partly caused by this. I know many beekeepers are using untested products such as thymol, wintergreen oil, and the insecticide neem without knowing whether they really work or how to apply them properly. I continue to be amazed at the many calls I get from American beekeepers asking me how to use these illegal products. Actually, their questions are not of the "how to use" variety, but rather more like "I used this product last season and lost most of my colonies. What did I do wrong, and how should I use it this year so that I don't lose colonies?"

Lost in the deluge of Internet rumors is the fact that wintergreen, other essential oils, and neem have not been proven to work in mite control. Incomplete and unsubstantiated studies suggest these products may deserve additional testing, but certainly this is not sufficient for beekeepers to trust their livelihood to these compounds. Essential oils have been tested by numerous scientists, and the data suggest they may work in some situations but be utter failures in others. Until we determine how to use these oils with consistent success, a beekeeper using them is playing beekeeping Russian roulette. The insecticide neem has also been tested with some success as a miticide, but only very preliminary data are available. Mode and timing of application, proper doses, analysis of residue levels, effects on honey bees, and testing under diverse beekeeping conditions must still be done before anyone should be recommending neem to beekeepers.

I might understand panicked beekeepers reaching into the cornucopia of illegal snake oil remedies if there were no mite control methods that worked. This is not the case; menthol, oil-based extender patties, and formic acid work just fine against tracheal mites, and Apistan and formic acid work against *Varroa* mites. I use these products as recommended with great success; if an "ivory tower" university researcher can control mites, certainly more experienced "real-world" beekeepers should be able to do it. I suspect that recent colony losses are due to improper applications of legal chemicals, cutting economic corners by skipping a spring or fall application, poor or no sampling to determine mite levels, or using unregistered products instead of the legal ones.

Application methods, dose, and timing are crucial to the success of mite controls, yet beekeepers trying unregistered products are gambling that they will hit the right combination for successful mite control. The plant-derived insecticide neem is a good example. Neem comes in many forms, including oils and various emulsifiable concentrates, and many of these formulations do not dissolve well in water. Many beekeepers apply neem obtained from nonbeekeeping formulations without realizing that it is not getting into their hives. Neem's success against insects is due in part to its being a repellent, preventing pest insects from feeding on crop plants. Neem also repels bees, so if you are successful at dissolving neem in your sugar syrup, it may deter your bees from feeding. Your bees not only won't get their mite medicine, they won't eat their sugar feed either, and might starve to death over the winter.

That's some of the information you won't get from the Internet reports about neem. You also won't hear that some of the preliminary methods reported for neem did not work when other researchers attempted to replicate them, or that neem purchased from one source in Asia is not as effective as neem from a different source. No, what you read about neem, thymol, or wintergreen oil is simply too preliminary, too vague, and too unreliable to base your livelihood on it. The only reason for using these products at this time is if you want to get out of beekeeping in a hurry.

Residues are another issue that doesn't get fired around the world on the illegal miticide Internet grapevine. Honey today is rigorously tested by commercial buyers, especially for export markets, and the technology for honey testing is so precise that even one part of illegal product to a billion parts of honey can be detected easily. I wouldn't worry so much about residues if I wasn't so concerned about a loss of consumer confidence in our honey. If you want to get your own load of honey rejected because of illegal residues, that's your problem. But it's only going to take one report of illegal miticide being found in honey to shake the confidence of consumers who believe honey to be a natural, pure product. All the good work of the National Honey Board to increase honey sales could disappear following one Internet-amplified report of illegal pesticides in honey. Think about it; would you put honey on your toast after reading a distorted scandal article about how beekeepers are dumping illegal toxic pesticides into their hives? No, we need to be hypercautious about our beekeeping practices in an age in which we rely on miticidal chemicals to keep us in business, and the information highway amplifies rumors until they become fact.

I support the role played by a strong research community in testing and evaluating new products, and in protecting beekeepers from illegal and improperly tested products. Don't get me wrong; I'm not telling beekeepers to butt out of the research world. Researchers and beekeepers need to work in tandem, and beekeepers have a vital role to play in directing research into areas useful for them, in working with researchers in large-scale commercial trials of potential products, and at some point in deciding a project's usefulness through their decisions to buy or not buy a product.

I question whether beekeeping is being well served by the vague reports of miticide success that whip around the world instantaneously with the touch of a computer key. Until the Internet comes up with a way for recipients of information to evaluate its credibility, I would be very hesitant to base my mite control or any other management on something I assume is reliable only because it pops up on my computer screen. Next time you surf the Web and find a report about miticides, ask yourself how many times it was replicated, what variables were tested, how rigorous were the statistics analyzing the data, or even whether any data at all are included with the claims of product efficacy. Ask yourself whether you're willing to risk your colonies on the sort of information you're looking at, and whether you are confident enough in the reputation of the senders to commit the health of your bees to their reports.

Finally, copy the report over to an extension agent or research scientist you know and trust and ask for an independent evaluation of the information. Any report should be able to withstand the rigor of close and independent examination about its methods and findings. Remember that Internet information is unfiltered by questions and unregulated by peer review. The best way to evaluate what you receive is to know and have confidence in the sender, and to view new information critically and rigorously before committing your business to its recommendations.

18. Bee Flu

IWAS CHANNEL SURFING THROUGH TV-LAND A FEW nights ago when my wandering, tired brain was snapped back to attention by grotesque images of dead people rotting in their cars inside the Lincoln Tunnel in New York City. The show turned out to be a Stephen King made-for-TV special in which a mysterious and lethal flu devastates the United States, killing most Americans and leaving only a few survivors to reestablish civilization. The swollen, decomposing bodies in their cars reminded me of something, and the next morning I realized where I had seen such a thing before: sacbrood-infected bee larvae dead in their cells. Change the flu virus to the sacbrood virus, make the show about a devastating new variety of this bee disease, have the new sacbrood wipe out most of our colonies, and you end up with a bona fide made-for-TV, Stephen King horror movie.

Human viruses, of course, are no joke. The virus that causes influenza, for example, appeared in a particularly lethal variety at the end of World War I and killed about twenty million people. The HIV virus that causes AIDS and now infects more than ten million people worldwide was virtually unknown only fifteen years ago. The worst thing about viruses is that there are few effective treatments for them. Viruses are not like bacteria, which generally can be defeated with antibiotics. In the case of viral infections, physicians can only treat the symptoms caused by the virus and hope the natural course of the infection completes its cycle with the patient still alive.

Bee viruses, in contrast, are not as serious as the worst human ones. Most bee diseases caused by viruses are more like the common cold than a deadly flu. They can cause some damage to the colony, and severe cases are known, but in general bee viruses are a low-level nuisance and usually are not high on beekeepers' disease-prevention priorities. The greatest danger from bee viruses lies in the future—not from the ones we know about today, but from as-yet-unevolved and possibly more virulent descendants of today's relatively benign viruses.

The word *virus* means "slimy liquid" or "poison," and the existence of viruses was not discovered until early this century. Viruses are neither plants, animals, nor bacteria; they occupy their own separate kingdom in the world of living organisms. These tiny organisms (the largest virus is about 0.000002 inches across) are among the most primitive of all living things. Viruses are also the ultimate parasites; they depend almost completely on the host to perform most of their life-sustaining functions. They cannot make proteins or generate energy, and force the host cell to perform these functions for them.

Basically, a virus is nothing more than a minimal amount of genetic material encased inside a membrane. When the virus finds a host cell, it injects its genetic material inside the host. The virus's genome inserts itself into the genes of the host, and thereby takes control of the host cell's activities. The viral command center then forces the host cell to create an ideal environment for viral growth and reproduction, usually resulting in cell death and the release of many more viruses that move through the infected organism searching for more host cells. Viruses can lie dormant within a host for years, and can be transmitted easily between hosts even in their nonsymptomatic stages.

The identified bee viruses have a bewildering array of names and symptoms. They include such interesting-sounding names as chronic paralysis, cloudy wing, Kashmir, Egypt, sacbrood, Arkansas, halfmoon, slow paralysis, black queen cell, and acute paralysis. Most larval and adult bees have a number of these viruses present in their bodies, but symptoms usually don't develop unless colonies are stressed. Symptoms of active viruses include larvae that liquefy, darken in color, and fail to pupate; and adults with bloated abdomens and dislocated wings that are unable to fly and can be found crawling on the ground in front of colonies. There are no known treatments for any of the bee viruses, but requeening often reduces or eliminates the problem, suggesting that simply changing the genetic background of the bees in a colony is an effective response to viral infections.

Bee viruses can be a nuisance, and occasionally cause serious problems, but by themselves don't seem serious enough to justify a major research effort to further elucidate how they are transmitted, affect bees, and can be treated. Recently, however, I have become aware of increasing interest among beekeepers in expanding viral research. Fear of introducing new viruses is beginning to have an impact on bee policy decisions such as importations of European stock into Canada and the United States for breeding purposes, and the potential opening of U.S. borders to importations from such countries as Australia and New Zealand. Further, U.S. and Canadian regulatory officials are considering conducting a broad survey of North

American bee viruses to determine which ones are currently present in North America.

This recent flurry of interest in viruses has not been stimulated by the current level of viral impact on bees, but rather by the fear that a new virus species or variety may appear in North America and could devastate our industry, much as HIV, which causes AIDS, is wreaking havoc among humans. Indeed, AIDS is an excellent example of how quickly a new virus can mutate from a nonvirulent form into a serious disease–causing organism. The AIDS virus, HIV, evolved anywhere from fifty to nine hundred years ago, depending on which research source you believe. There are at least fourteen identified human and monkey viral relatives of HIV, none of which is virulent, and all of which probably move freely between humans and monkeys. For some reason HIV mutated into the deadly form that now infects humans.

How likely is this scenario to occur in honey bees? The greatest viral danger to our bees may reside within the closely related Asian honey bee species. Much like monkey HIV in humans, a nonvirulent virus in an Asian honey bee could mutate slightly and become highly virulent when hosted by *Apis mellifera*. We are all familiar with a similar incident involving an Asian mite. Remember *Varroa*, the mite that causes almost no problems for its Asian host *Apis cerana*, but is highly destructive in our colonies of *Apis mellifera*?

Viruses cause more damage in weakened colonies than in healthy ones, and the combination of *Varroa* and tracheal mites is providing plenty of weak colonies for viruses to grow in. Beekeepers in my area, the Fraser Valley region of British Columbia, experienced average colony losses of 66 percent in the early 1990s as a result of infestations of the two mites in conjunction with an unusually bad summer. Was this extraordinary colony mortality due only to mites and weather? I think not. It seems likely to me that another factor was at work here as well, possibly bacterial or viral infections that may have provided the final bullet to finish off colonies weakened by mites and a rainy summer.

There is some evidence that this scenario is plausible. The results of preliminary studies in American laboratories indicate that mite-infested bees have higher levels of bacterial infestations, and antibiotics administered as part of mite treatments seem to improve the treatments' effectiveness. If bacterial infections are higher in mite-infested colonies, then viruses likely are too, and may be contributing to colony death. In addition to producing a weakened colony in which pathogens can thrive, *Varroa* and tracheal mites may also be vectors transmitting viruses and bacteria between colonies.

Unfortunately, there is very little we can do to prevent the evolution of new and more virulent strains of bee viruses. The only effective way to control a new bee virus would be to destroy infected colonies and hope the virus had not spread, but the current dependence of the North American beekeeping industry on migration would make it difficult to effectively isolate and destroy a new viral pathogen. We could reduce the potential for outside viruses coming to North America by limiting bee importations, but there is no evidence that there are viruses elsewhere in the world that we don't already have here. Also, the current trend in both Canada and the United States is toward increased importation, particularly from New Zealand and possibly Australia. It seems heavy-handed to ban importation on the basis of a hypothetical virus that may or may not ever appear, and is just as likely to appear first in North America as elsewhere.

I think we're left with the Stephen King horror movie scenario: a virus *could* evolve that *might* wreak havoc in our industry. Fortunately, this possibility seems remote, but should it occur, there is very little we could do about it. I think we should at least begin a regular monitoring program for viruses to get some sense of what is currently out there. That might enable us to quickly pick up changes that might pose more severe problems than those we see in the current viral environment. We also can look to modern medicine and molecular biology, and monitor their increasingly sophisticated ability to understand and possibly eventually control viral pathogens. If a prevention or cure for the human AIDS virus is ever found, then treatment for a new bee virus would not be far behind. I hope we make progress quickly; I hate Stephen King movies.

PART THREE
Industry Politics

Industry Politics
Beekeeping is a culture unto itself. Although it is woven into the fabric of North American society, nonpractitioners are unaware of its extent or its rituals. That is not unusual; those of us who do not skydive, raise pigs, or play bridge have little contact with or knowledge about skydiving clubs, hog-farming organizations, or bridge tournaments. The general public occasionally hears about killer bees, reads an advertisement for honey-flavored cereal, or giggles at a picture of a beekeeper wearing a bee beard at a local meeting, but otherwise has little formal dialogue with those of us who keep bees.

Yet, beekeeping plays an enormous role in agriculture and as a hobby, and has its own set of organizations, politics, and issues. Beekeepers lobby government for handouts, complain about cheap imported honey, and attend an endless series of local, state, and federal meetings to talk about bees and matters associated with beekeeping.

Statistics about bees and honey provide a glimpse into the extent of beekeeping in North America. In 1996, 198 million pounds of honey was produced in the United States from 2.57 million colonies, with an average yield of 77.2 pounds per colony. The total price paid to American beekeepers for their honey was $177 million, at an average price of $0.894 per pound. The largest number of colonies in any state was 390,000, in California, followed by 240,000 each in Florida and South Dakota; there were only 6,000 colonies in Connecticut, Delaware, Maine, New Hampshire, and Rhode Island combined (there are so few commercial beekeepers in those states that data about colony numbers and honey production are not published separately to avoid disclosing information about individual operations). California and Florida produced the most honey, tied at $22,932,000 worth each, while the five states with the lowest colony numbers together produced only $288,000 worth of honey.

In Canada, the industry hovers between 600- and 650,000 colonies each year, with the highest concentrations in the prairie provinces of Alberta, Saskatchewan, and Manitoba. Canadian hives produce $50 to $70 million worth of honey each year, and colonies from some regions in the prairies consistently show the world's leading per-colony honey production figures. It is not unusual for a single hive on the Canadian prairie to produce more than 300 pounds of surplus honey during the short northern growing season. These high yields are due to extensive cropland planted in good honey-yielding crops such as canola and clover, and also to the long summer days during which bees can forage for sixteen to twenty hours each day.

Beekeepers are as social as their bees, and have organized themselves into organizations to represent their interests. Canada has two major beekeeping groups: the Canadian Honey Council (CHC), made up primarily of beekeepers, and the Canadian Association of Professional Apiculturists (CAPA), consisting of research and extension personnel who are employed professionally as government or university research scientists or hold extension or regulatory positions with our provincial and federal governments. The two groups meet together every January and share educational lectures, reports from various CHC/CAPA joint committees that work together throughout the year, and presentations by federal officials.

The situation in the United States is more complicated and chaotic. There are numerous organizations that represent only limited components of the industry. As the various organizations generally do not meet together, it is difficult for them to develop a common perspective on any issue. For example, bee researchers attend two major annual meetings, the Bee Research Conference held in October somewhere in the southern United States, and the larger Entomological Society of America meetings held in December at a different location each year. Beekeepers and extension personnel generally do not attend either meeting in any appreciable numbers. The "business" organization of the research community is the American Association of Professional Apiculturists (AAPA), which is similar to the Canadian association but lacks the regulatory component. The regulatory personnel meet annually at the Apiary Inspectors of

America (AIA) meeting. The AIA meetings focus on technical and political aspects of pest and disease management, and are attended primarily by government officials who regulate beekeeping. Sometimes the AIA and the AAPA may meet jointly, but such arrangements have been sporadic and inconsistent.

The last component of the U.S. industry's organizational structure is the most fractured. There are two major groups of beekeepers' organizations in the United States: the American Beekeepers Federation (ABF) and the American Honey Producers (AHP). These organizations split from a single society many years ago as a result of bitter personal and professional differences between some of the members, and have not succeeded in merging again in spite of recent efforts to explore reunification. The two groups represent similar constituencies and have similar meeting agendas, but meet at different times and in different places each year.

The national-level organizations don't even begin to reveal the complex nature of honey bee politics, however. There are innumerable local and state or provincial organizations that represent beekeepers, each with its own newsletter, meeting schedule, and perspective. For instance, there are about ten local clubs in and around my home city, Vancouver, British Columbia. Some are subsections of the provincial organization, the British Columbia Honey Producers Association (BCHPA), but other clubs refuse to join the BCHPA because of a real or imagined slight that took place years ago that no one can remember, or simply an unwillingness to pay the extra ten dollars per member it would cost to join. Each group has a particular character; one club is made up primarily of non-English speaking immigrants mostly over the age of eighty, another is extraordinarily active at bringing bees into the public schools, and a third serves as a central purchaser for antibiotics, equipment, and books for its members. And each has a different opinion about the beekeeping issues of the day.

In view of the local, regional, and national diversity in our beekeeping organizations, it is not surprising that both the American and Canadian beekeeping infrastructures have had difficulty dealing with political issues in recent years. This fractured organizational structure has impeded lobby efforts and has made it difficult for beekeepers to present a cohesive voice to the legislative bodies that ultimately determine the fate of beekeeping-related issues. This has caused some damage to the industry. For example, the research and extension communities in both countries have been losing university and government positions steadily during the last two decades because the beekeeping industry has been ineffective at lobbying to maintain much-needed expertise in management research, delivery of extension information, and regulatory procedures.

New and well-coordinated initiatives to deal with problems such as mite parasites have also been lacking, partly because there are not enough employees to work on these issues, but also because it has been difficult to coordinate programs. There have been innumerable programs to select disease-resistant queens throughout North America, with considerable overlap and duplication of effort epidemic in them. In this time of reduced resources and diminishing funding, we simply cannot afford to run carbon-copy versions of research projects at a number of laboratories, yet this is exactly what we have been seeing in North American bee research. Similarly, a centralized effort to select, breed, and rear queens would be singularly useful, yet the fierce independence of the beekeepers who buy and sell queens precludes any such effort from success. Honey quality is another issue. Instances of honey adulteration with corn syrup continue to flare up in our two countries, yet there is no central, commonly used laboratory available to beekeepers and honey packers to test samples for adulteration.

International issues also create considerable dissent within and between countries. Canada closed its border to bees from the mainland United States in 1988, creating animosity between American and Canadian beekeepers that persists to this day. This importation quarantine was instituted because tracheal mites and then *Varroa* arrived in the States, and was clearly in Canada's interest. Yet, even within Canada there was some dissent, and the federal government had difficulty determining whether the federal-level Canadian Honey Council or local organizations from the same district as the minister of agriculture—who thus had his ear—were the proper representatives of the beekeeping industry.

Perhaps the most surprising aspect of bee politics is that beekeepers do not always express the lessons of cooperation and cohesiveness that their bees manifest day after day. Beekeepers are highly social and interactive, like their bees, but we are not nearly as effective at establishing the chain of command, understanding our individual roles, interpreting our responsibilities to work toward a common purpose, and maintaining a collegial, constructive, and well-functioning social structure. We should remember that bees have been around much longer than we have, and learn from them. Perhaps there is hope that, as we evolve and develop as a species, beekeeping politics will follow the lead of the bees and mature into a social structure as enviable as that of the hive.

19. Finding Dirty Honey

THE SUCCESS OF THE BEEKEEPING INDUSTRY IS based on the taste, texture, aroma, and image of honey. Today's food shopper is obsessed with nutrition and health, and honey's image as a pure natural food is highly attractive to consumers. The reality is that honey is mostly sugar, and expensive sugar at that, but beekeepers have been successful at marketing honey because of its many desirable flavors and the public perception of honey as a somewhat mystical and healthful food product. Indeed, if beekeepers were included in one of those "what profession do you most respect" polls, we would be right near the top for providing consumers with this clean, pure commodity.

Clean and pure from the consumer's point of view, that is. From a chemist's perspective, honey is a dirty product full of hundreds of compounds and contaminants—natural and added by the beekeeper—and a nightmare to analyze. Luckily, most of the trace compounds found in honey are floral in origin or produced by the bees, and it is these minor substances that give honey its various tastes and aromas. The purity of honey is increasingly threatened by poor beekeeping practices that contaminate honey with repellents, pesticides, and antibiotics, however, and significant quantities of honey that have been adulterated with artificial sweeteners continue to turn up.

Most of us are blissfully unaware of the extensive food analysis subculture that keeps our food products pure, including honey analysis. I wrote to just a few of the experts in this field to get some background material on how contaminants are identified in honey, and was deluged with reprints and articles that describe the various analytical techniques. Food chemists have devised clever methods to detect hundreds of compounds in honey, from the natural sugars and aromatic compounds that belong in honey to pesticides that have no business there. The techniques are elegant and chemically sophisticated, and involve numerous methods to separate compounds and identify them by their chemical "signatures." Chemists can identify as little as one part per billion of many compounds. One chemist described his analytical capabilities to me by saying that he could find a level of contaminant equivalent to one grain of sand on a beach! Considering the sophisticated nature of these analytical techniques, it's a wonder any honey passes inspection. Honey is often considered to be contaminated if it contains one part in ten million of an unregistered agricultural chemical.

Today, numerous laboratories worldwide routinely examine honey because contamination with beekeeping products is a significant problem, and also because honey adulterated with sweeteners continues to surface in national and international markets. Honey analysis is not as simple as it might appear, however. Sophisticated tests to discover minute quantities of compounds in honey have also revealed that many substances once considered contaminants are actually "natural," produced in floral nectars or by bees in processing the nectar into honey.

Formic acid and menthol are good examples of compounds used in beekeeping that also occur naturally. Canadian beekeepers wanted to register formic acid for use against tracheal mites and *Varroa*, and preliminary tests indicated that it was effective when properly used. However, formic acid residue levels of about two parts per million—about twenty times higher than the tolerance level—were being found in honey from treated colonies. There was some concern about this until honey from untreated colonies was analyzed and found to have natural levels of about one to three parts per million of formic acid. In other words, the apparent residue levels in honey from treated colonies were no higher than natural levels. Colonies treated during the summer honeyflow showed formic acid levels of about eight parts per million, approximately four times higher than the natural levels. These results had two important implications. First, Canadian beekeepers were able to register formic acid for use in Canada, because proper applications did not result in residue levels higher than naturally occurring amounts of formic acid. Second, the results emphasized the importance of not applying formic acid during honeyflows, because summer applications resulted in residue levels above those normally found in honey.

Menthol is another compound used in mite control that occurs naturally in honey. Natural menthol levels are usually at or below our one part in ten million tolerance level, but honey from treated colonies may have menthol levels of one to two parts per million, about ten to twenty times higher than the Canadian standard for unregistered chemicals. Further, menthol in honey persists during storage, and so this high level could pose some problems according to standards used for agricultural chemicals. Menthol is undetectable to consumers at levels below about thirty-six parts per million, however, so even levels of one or two parts per million would not affect the honey's taste. Further, the beekeeping industry argued successfully that menthol's status as an already approved food additive should give it a higher tolerance standard than that typically used for agricultural chemicals, and so menthol use is permitted in Canada, as it is in the United States. Treatments immediately preceding and during the honeyflow are not recommended because they can result in unacceptably high menthol residues in the honey.

Bee repellents such as phenol (carbolic acid) provide another good example of how our increasing capability to detect foreign substances in honey has affected beekeeping practice. Beekeepers use repellents to drive bees out of honey supers before bringing them in for honey extraction. Phenol is not licensed for use in Canada or the United States, and the maximum allowable levels of phenol in honey are low (one and five parts in ten million for Canada and Germany, respectively, and zero in the United States). Yet, a 1985 survey found that forty of sixty-seven honey samples contained phenol levels from one to eleven parts per million, with average values of five parts per million. Another study found that phenol levels in honey diminish over time, so the high levels found in those samples are best explained by phenol use that year. These chemical analyses led to increased regulatory vigilance in Canada and the United States, and illegal use of phenol has been decreased.

New techniques for detecting the antibiotics fumagillin (Fumidil), oxytetracycline (Terramycin), and sulphathiazole have also resulted in increased regulatory activity. The presence of antibiotic residues is of concern both because they may cause allergic reactions in consumers and also because the organisms responsible for the foulbrood and nosema diseases targeted by these drugs could become resistant if overexposed to medication. Thus, most countries have zero or very low tolerances for antibiotic residues, and require that imported honey be completely free of these substances. Sulphathiazole is an unregistered drug almost everywhere in the world, so its presence in honey is particularly troublesome. Honey is routinely tested for these three antibiotics, especially imported honey, and loads are rejected when even trace residues of any of them are found. Surprisingly, sulphathiazole residues still appear in honey, although it has been illegal to use this drug for well over a decade.

Acaricide contamination following treatments for tracheal mites and *Varroa* is the most recent honey residue problem. Unacceptable levels of legal compounds such as fluvalinate occur when hives are treated close to or during the honeyflow, and there is some concern that fluvalinate accumulating in wax after repeated treatments could lead to honey contamination. Although proper use of fluvalinate does not result in contaminated honey, years of its use even according to the label recommendations can result in wax contamination. We are beginning to see loads of beeswax rejected for importation because of fluvalinate residues, and it is becoming increasingly difficult to find clean, pure wax for the international trade.

Of even greater concern is the appearance of illegal, unlicensed agricultural chemicals in honey. In 1995, for example, both the United States and Canada rejected honey from China because it was contaminated with chlordimeform, an acaricide previously used against mites on crops and ticks on cattle, among other things. This chemical is banned in most countries because it has been linked to high levels of bladder cancer. The presence of chlordimeform in the honey was alarming, as it indicated that Chinese beekeepers were using a banned, dangerous substance to combat tracheal mite and *Varroa* infestations. The only good news is that loads of such contaminated honey are refused entry into North America, which relieves some of the economic pressures caused by cheap imported honey!

A final adulteration problem is the addition of sweeteners to honey. The relatively low price of commercially available sugars tempts unscrupulous producers, importers, and packers to extend the volume of honey sold by adding sweeteners. High fructose corn syrup (HFCS) has been a particular problem because its sugar profile is very similar to honey's, and is thus difficult to detect. For example, pure citrus and mesquite honeys can test as being adulterated with corn syrup when their sugar profiles alone are analyzed. Recent advances in honey analysis have considerably improved our ability to detect corn syrup adulteration. Interestingly, the techniques used today do not test for sugars, but rather for minute quantities of proteins found in honey and corn syrup that can be used to differentiate the two substances with great accuracy.

Given our ability to test for sweetener adulteration and the common use of these tests in both national and international markets, it is surprising that loads of adulterated honey still appear on the market. A major honey packer in Mississippi was convicted in 1996 of adulterating millions of dollars worth of honey, and both U.S. and Canadian officials continue to find and reject loads of honey from overseas that are heavily contaminated with corn syrup. Every sample from one large shipment of honey in 1995 was found to be adulterated at some level, and the honey from one particular honey packer contained 47 percent corn syrup! This shipment was first rejected by the United States and then made its way by a circuitous international route to Canada, where it was discovered and rejected again.

The take-home message here is clear, but the continued presence of contaminants in some commercial honey indicates that not everyone has gotten it. Honey's sales depend on its reputation as a pure product; improper use of licensed chemicals, the use of unlicensed chemicals, and the adulteration of honey with cheap sweeteners threatens that reputation. If you don't care about the industry as a whole, have little regard for the purity of your product, and are a maverick beekeeper who disregards regulations, then consider this: adulterate and contaminate your honey, and you're very likely to get caught. Even minute quantities of contaminants are detectable, and regulatory officials have become increasingly aggressive at sampling all our food products, especially honey. We all need to remember that honey is a marvelous product and as natural a substance as any food on the market today. Let's keep it that way.

20. Border Closure

ON JANUARY 1, 1988, CANADA PROHIBITED ALL IMportation of honey bees from the United States. The ban was enacted following the discovery of *Varroa* mites in the United States, and followed increasingly strict sets of import regulations enacted since 1984 to prevent the importation of tracheal mites into Canada from the United States. The initial 1988 embargo on bee importations was for two years, but the ban has been renewed every two years since its initial enactment. Hawaii has been exempt from the quarantine regulations since 1992, after extensive surveys failed to discover either *Varroa* or tracheal mites anywhere on the Hawaiian Islands. The quarantine against bees from the mainland

United States will remain in place at least until December 31, 1999, at which time it will either be renewed or modified to permit some importations. Canadian beekeepers discussed their views on reopening the border at the annual Canadian Honey Council meetings in January 1997. This meeting was an interesting one, and its participants both confirmed the wisdom of the initial border closures and supported continuing the ban on bees from the continental United States.

At the risk of offending American readers, I must say that I believe the border closure was the appropriate action, and continues to be justifiable because it has protected Canadian beekeepers from both tracheal and *Varroa* mites. The objective of quarantines is to protect the majority interests from medical or agricultural damage from foreign pests or diseases. Quarantines are not necessarily designed to prevent pest entry forever, although some have that objective. Instead they generally attempt to slow the progress of a potentially harmful pest as long as possible, or at least as long as the quarantine is economically preferable to the damage the pest may cause. In both objectives—economic protection for the majority of Canadian beekeepers and slowing the spread of tracheal and *Varroa* mites—the Canadian ban on bee importations from the United States has been a success.

Many American beekeepers do not realize that the embargo was initiated with broad support from Canadian beekeepers. There seems to be some feeling among U.S. beekeepers that a small cadre of misinformed government and university personnel enacted the quarantine against the wishes of most Canadian beekeepers. This was not the case; almost every Canadian provincial beekeeping organization has lobbied actively in favor of the quarantine since the original 1988 closure, and the national beekeeping organization, the Canadian Honey Council, effectively lobbied to initiate and then continue the border closure based on that broad provincial support. There has been some opposition, particularly from beekeepers in the Peace River area of British Columbia and Alberta who prefer package bee systems, but I think it fair to say that most beekeepers in Canada, and certainly their officially constituted organizations, strongly support the border closure.

There have also been misconceptions concerning the objectives of Canada's embargo. No one in Canada believed then, or believe now, that the embargo would prevent the arrival of either *Varroa* or tracheal mites. Rather, the border closure was designed to slow the spread of these mites for as long as possible, for two reasons. First, the pesticide-licensing process in Canada is considerably more involved than in the United States, and in 1988 Canada beekeepers had no registered chemicals to use against either mite. It was only in 1994 that fluvalinate was registered for use in Canada, finally providing beekeepers with some chemical protection against *Varroa*.

Second, Canadian beekeepers believed we could slow the mites' spread to such an extent that widespread, expensive treatments could be delayed for many years, a belief the ensuing years have proven correct. In 1993, tracheal mites were present in only 16 percent of Canadian colonies, and *Varroa* mites were in fewer than 1 percent, and those in a very few locations along the U.S.-Canadian border. In 1997, fewer than 25 percent of Canadian colonies in most provinces had *Varroa*, and some major beekeeping provinces such as Saskatchewan were almost *Varroa*-free. Savings in treatment costs alone have exceeded three million dollars annually, well worth any negative impact caused by the embargo.

The border closure also benefited Canadian beekeepers in other ways. We have become more self-sufficient in our beekeeping operations, although that was *not* a primary objective of the embargo. Many beekeepers have found overwintering systems to be more cost-effective than the old package bee management methods. In addition, queen rearing, package bee production, and sales of nuclei are providing new sources of income for Canadian beekeepers, and the diversity of new management systems generated by the need to be self-sufficient has been healthy for the industry.

Nevertheless, the quarantine has had adverse effects on some Canadian beekeepers. Beekeepers in the northern beekeeping regions of Alberta were the most seriously affected, because almost all of them depended on package importations. A number of beekeepers throughout Canada went out of business in the late 1980s, primarily due to low honey prices, although the sudden embargo on package importations was a contributing factor in reducing the number of Canadian colonies from about 700,000 to 520,000 in 1992. Today, however, our colony numbers have climbed back up to around 650,000, again due more to the increased price of honey than any other factor.

American beekeepers were far more seriously affected by the ban, and the embargo clearly has not been in their best interests. Before 1988, Canadians purchased more than 300,000 packages a year from U.S. beekeepers, mostly from California but also from other areas in the southern United States. Annual package and queen sales to Canada amounted to more than twelve million dollars when the border was closed, and loss of the Canadian market was a major blow to the U.S. industry. More than just sales were lost, however; many U.S. and Canadian beekeepers had developed close friendships, even marriages between families, and a number of U.S. and Canadian beekeeping operations had close economic ties. Package bees were a way of life for many, and the change to new systems proved to be a difficult adjustment for beekeepers on both sides of the border.

The long-term status of the border closure is not clear at this point, and *Varroa* and tracheal mites are becoming more widespread. The arguments favoring border closure today focus on two issues: continued restriction of mite spread and stopping the entry of Africanized bees. The arguments for opening the border balance these negative aspects with the economic advantage of cheap and easy access to spring queens and packages, although it is unlikely that package importations will ever approach their former levels.

The effectiveness of a continued quarantine in slowing the further spread of mites in Canada is a judgment call; at some point it will no longer be justifiable to prevent importations on that basis, although the precise date when that will happen is hard to predict. Certainly the registration of fluvalinate and formic acid in Canada has changed the balance considerably. Nevertheless, beekeepers from noninfested regions still support the border closure. The beekeeping industry is much less mobile in Canada than in the United States, so the spread of mites can be contained more effectively for a longer time than was possible in the United States.

A more serious problem blocking renewed importing of queens or packages into Canada will be Africanized bees, which may surprise some American beekeepers. Our northern climate will prevent these bees from arriving here on their own, and many Canadian beekeepers wonder why we would risk importing them, particularly since our industry is doing quite well without any bees from the United States. Canadians have taken a "wait-and-see" attitude on this issue, because it is not yet clear how problematic these bees will be in the United States and how successful queen rearers in the southern states will be in preventing matings with feral Africanized bees. In any event, bees imported into Canada from the United States will require strict certification that the queens are not Africanized, which may make such importation economically prohibitive. Early spring queens imported from Australia, New Zealand, and Hawaii, or late spring and summer queens reared in Canada will be far cheaper. Thus, opening the border to U.S. bees probably will not produce a huge influx of queens and packages, at least initially.

Whatever your opinion on the border closure issue, it is important to remember that past decisions to close the border have had wide support in Canada, and have turned out to be in the best interests of the majority of Canadian beekeepers. The situation is slowly changing, however, and at some point the embargo on U.S. bees may be relaxed. Even if the border is reopened, importation will involve costly inspection procedures to certify that imported bees are disease-free and non-Africanized. We won't see a wholesale return to package bee systems, except possibly in the Peace River region, and even there many beekeepers may choose to overwinter the majority of their colonies. Correctly or not, mites and Africanized bees have changed the fabric of Canadian and American beekeeping forever, and looking to the past will not alter that fact. In the future, Canadian beekeeping will be largely self-sufficient. The package way of life is gone, and much as we may wax nostalgic about it, I don't think we will see many package bees in Canada again.

21. Government, Queens, and Brother Adam

I NEVER MET BROTHER ADAM, CORRESPONDED with him, or even used his Buckfast queens. Yet, his death in September 1996 touched me, partly because of his reputation but also because of an image from a television show about him that was broadcast a few years ago.

If you don't already know, Brother Adam was a Benedictine monk and bee breeder from Buckfast Abbey in England who developed a line of bees called the Buckfast bee. He was a remarkable individual whose knowledge of breeding was intuitive rather than the result of scientific training, and the techniques he developed to select and maintain queen stock have become classic methods. The television image of Brother Adam that came to mind when I heard of his death was of him at the age of ninety being carried in a bamboo chair up Mount Kilimanjaro in Africa, still searching for new bee stock in spite of his advanced age. He explored the remotest regions of the world looking for qualities to mingle with his Buckfast stock, and even into his nineties he continued to try and improve his already stellar queens.

One of the things that impressed me about Brother Adam was the fact that he worked for an "organization," yet was able to maintain and sell a line of bees for many decades. This is a singular achievement, because other organizations such as universities and government laboratories have not been successful at keeping and selling breeding stock for commercial beekeeping. Beekeeping is in many ways an odd agricultural profession, and one of its oddities is the failure of public organizations to maintain selected and commercially viable breeding stock.

This is not to say that universities and government have not been involved in stock selection. On the contrary, there have been innumerable programs to select stock with characteristics useful to beekeepers. The problem, however, is that programs to select stock are run on government time, meaning that funding is provided for short-term (three to five years) selection but not for long-term maintenance.

A good example of this problem is the British Columbia bee. In the late 1970s and early 1980s, the B.C. Ministry of Agriculture funded a program to select, test, and breed queen stock with qualities especially desirable for British Columbia conditions. This program was conceived and headed by John Corner, the provincial apiculturist at the time, and after three years did produce an excellent bee. Corner's group followed the Page/Laidlaw closed mating system, a breeding pattern developed by Rob Page and Harry Laidlaw that allows open mating by a limited number of queen lines to be maintained almost indefinitely without inbreeding. It is a system used today by many commercial queen rearers, and if I remember correctly the B.C. bee researchers were the first to demonstrate that this system could be effective for stock selection and maintenance on a large-scale, commercial basis.

Colonies headed by the B.C. queens had all the qualities you could ask for in a queen. They were gentle, overwintered well, swarmed rarely, and produced buckets and buckets of honey. Queens from this line also laid the finest brood patterns I've ever seen, a stunning array of cell after cell, frame after frame, filled with uniformly healthy and solid brood, a beekeeper's dream.

Yet, when the project ended, the stock gradually disappeared. Breeder queens were sold to private commercial beekeepers and gradually became absorbed into the general queen population. Remnants of this stock remain in British Columbia; the queens we use at Simon Fraser University are descended from the original B.C. bee. It would be a stretch, however, to say that anyone in British Columbia or elsewhere is still selling the pure B.C. bee.

The gradual disappearance of the B.C. bee does not mean the program was worthless, of course. We now have a viable and growing queen-rearing industry in British Columbia whose history can be traced directly to the B.C. bee. The breeding program was only one component of a broader project to train beekeepers to select and rear queens. The individuals who worked for or took courses from this program formed an experienced nucleus of commercial queen rearers who have continued in the industry. Thus, although the B.C. bee is hard to find, the B.C. queen industry continues to thrive as a result of the program.

The history of the British Columbia bee is typical of stock selection programs. I have seen the same process repeated again and again. A scientist succeeds at convincing a government agency to provide three to five years of funding to select "improved" queen stock, but at the end of the program no funds are provided to maintain the stock and it slowly vanishes as an independent entity. I won't embarrass anyone outside my home province by listing all the programs that have ended this way, but memories of selected and lost stock are strewn all over the North American landscape.

I perceived this problem during the B.C. bee episode, when I was a young and naive new faculty member at Simon Fraser University. I recall attending a meeting of the Canadian Association of Professional Apiculturists in the early 1980s and suggesting that we put together a proposal to develop a Canadian Queen Stock Center that would maintain selected stock for sale as breeder queens to commercial beekeepers. My more experienced colleagues sitting around the table were polite, and humored me by agreeing that it was indeed a good idea, and I should develop a preliminary proposal with a budget. My youthful enthusiasm carried me as far as the budget, which reached into the millions for initial start-up funds and continued on at hundreds of thousands of dollars each year for operating funds. Even my starry eyes clouded over when I saw the costs, and my proposal vanished into well-deserved obscurity.

But then there's Brother Adam, an individual who succeeded in selecting and maintaining a particular stock for decades while other bee stocks have disappeared within a few short years. This achievement is not only notable in itself but has also turned out to be important for North American beekeeping, because the Buckfast stock has been incorporated into tracheal mite–resistant stock now being sold commercially in Canada and the United States. Why did Brother Adam succeed where so many other programs failed?

One part of the answer is that he had the long-term financial support necessary to maintain stock. Buckfast Abbey was incredibly supportive of Brother Adam, and funded his work on the long-term scale necessary to maintain stock indefinitely. Granted, a government laboratory is not a monastery, but nevertheless, the Buckfast example demonstrates that stock maintenance is possible if funding is committed to that objective.

Brother Adam also succeeded because he was a hard-nosed businessman. He sold his breeder queens at a good price, and developed licensing agreements worldwide so that he continued to receive royalties on queens derived from his stock. Thus, Buckfast Abbey tried to operate as a commercial rather than a public organization, and did not consider profit a dirty word.

A final aspect contributing to Brother Adam's success was the product itself, considered by many the most prolific, disease-resistant, and honey-producing queen ever bred. Whether the Buckfast queens are indeed the top of the line is arguable; I'm sure there are other queens that do as well or better in particular parts of the world. Nevertheless, Brother Adam's queens certainly are among the best queens ever bred, reared, and sold around the world, and their reputation contributed to the ability of Buckfast Abbey to maintain the stock.

Perhaps it's time to think again about stock maintenance in North America, and about what we should do with the stock bred in the selection programs that have littered our beekeeping landscape in the last few years. For example, it is hard to keep track of all the government-funded projects that have been conducted recently to develop mite-resistant stock, and I am concerned that any benefits that might be derived from these mite-resistant queens will dissipate when the programs end.

We can duplicate the success of Brother Adam, but only if a new model for stock maintenance is developed. Clearly, a stock center must be financially independent, meaning that proceeds from the sale of breeder stock would have to be sufficient to cover the center's expenses. Since government has proven unsuccessful at long-term programming for the bee industry, I suggest that commercial beekeepers might set up a stock center themselves. If all the queen breeders in North America raised the price of their queens by ten cents each and contributed that amount toward the establishment of a stock center, it would take only a year or two to raise enough start-up funds to get it going. Operating funds might be generated for a few years by a continued levy, but over time the levy should decrease and the center become self-sufficient.

I propose this concept with no expectation that it will happen. Beekeeping is a competitive commercial enterprise, and it seems unlikely that an industry composed of highly independent individualists would be able to cooperate to the extent necessary for a stock center to succeed. Reluctantly, I accept the fact that Brother Adam was unique, and the circumstances under which he worked unusual. I doubt we will see another enterprise like Buckfast Abbey and the Buckfast queens for a long time. I didn't know Brother Adam, but I'm going to miss him.

22. Positions

I RECEIVED MY PH.D. FROM THE UNIVERSITY OF Kansas in 1978, and began the expected search for a position. My hope was to land a job as either a university professor or a research scientist with the U.S. Department of Agriculture. Then, as now, there were few Ph.D.-level jobs available in pure apiculture, so

I applied for positions in related fields such as animal behavior, ecology, behavioral ecology, and evolutionary biology.

At first, I treated the search process as a bit of a lark, since I was confident of landing a job. I decided to demonstrate my disdain for any employer who would reject me by taping what I expected to be a very few incoming rejection letters on my office wall. This turned out to be a bad idea; my spirits sagged as the letters multiplied from a patchwork pattern into solid wallpaper. At one point I counted more than fifty rejections staring down at me from all sides. The last straw was a letter from Yale University rejecting me for a job *I had not even applied for*! That day, I tore down the notices from my walls, and for the first time began to think that the mythical position I had spent so many years in school training for might never materialize.

I did, of course, eventually get a job, first teaching for a year at Idaho State University, then landing my current faculty position at Simon Fraser University (SFU). This job was advertised in the field of "insect pest management," an area I knew nothing about, but the Biology Department decided that killer bees were close enough to being pests for their purposes and hired me. I was fortunate that the department encouraged me to build a research program in apiculture at SFU. However, most of my teaching responsibilities were, and remain, in nonapicultural areas. For example, I teach Introductory Biology, Introductory Entomology, Pest Management, and Chemical Ecology, all subjects that might involve bees but certainly don't focus on them.

I considered myself an apicultural rebel when I was hired, because my doctoral training was not purely in beekeeping. Rather, I was a behavioral ecologist who happened to work with bees. This was unusual for a bee guy in the late 1970s; most people who studied bees back then came from a more traditional background. In fact, there were quite a few of us using honey bees for more general research in genetics, physiology, behavior, ecology, and so on. We thought of ourselves as the new wave in bee academia and considered the older generation outdated and narrow.

It is, of course, ironic that I am writing this book today, because I have become the generation I used to complain about. I now find myself among the few university faculty who still practice the craft of practical bee management research. Indeed, I have become increasingly alarmed at how little bee-related jobs have to do with beekeeping. The pendulum in research and academic positions has swung too far away from bee management. The apiculture research community is losing the capability to conduct the beekeeping and pollination research that contributed so much to our industry in previous decades.

Take universities, for instance. It is extraordinarily difficult to get a job as a professor in any field, let alone apiculture. Each advertised position gets hundreds of applicants, of whom at least forty or fifty turn out to be viable candidates. The list is narrowed down to four or five individuals who are brought in for interviews, during which each must give one or two seminars and talk one-on-one and in small groups with almost everyone in the department. Then the department members meet to vote on their choice, which must be approved by deans and vice presidents before becoming official.

The problem with bee jobs today is university politics, not the quality of the candidates. I can think of at least ten highly qualified recent graduates who could fill a position involving teaching, extension, and research responsibilities with honey bees. However, departments and deans can and do take jobs advertised in apiculture and change them around to fill a different mandate. The hidden agenda in contemporary job advertisements is often for a biochemist or molecular biologist, today's trendy "buzzword" fields.

A typical job-search scenario might begin when the incumbent retires or when beekeepers convince a university to create a new position in apiculture. The job is advertised, numerous candidates apply, and a few are invited to interview. The short list usually includes some candidates who study beekeeping or pollination management as well as others who happen to work with bees but whose focus is more academic than practical. Departmental opinions are split between those who want the beekeeping candidate and those who don't give a hoot about beekeeping and want someone in a different field that complements their own research. More often than not, the department picks the academic over the practical researcher, or the dean insists on hiring the academic candidate, and another bee job is lost.

Positions with a significant beekeeping component are disappearing from the U.S. Department of Agriculture (USDA) as well, but for different reasons. For one thing, the federal government is closing its bee laboratories. The Wisconsin and Wyoming laboratories have already been shut down, and now the Tucson lab may be toast as well. Although some positions are shifted to other laboratories when this happens, the net result is a loss in jobs.

Sometimes USDA bee laboratories are used as destinations for non–bee researchers when other government laboratories are closed. A USDA lab in some unrelated field is shut down, but senior personnel have to be parked somewhere until their thirty years of employment are completed and their pensions kick in. Sometimes, these nonapicultural researchers are moved to bee laboratories to eke out their remaining time, occupying a position that could be filled by a younger, more bee-friendly person. Some of these transferred scientists have made valuable contributions to bee research, and the outsider perspective has sometimes proven highly valuable to the beekeeping industry. Nevertheless, although occasional transfers of this nature are healthy, the frequency with which transfers have occurred in recent years, and the degree to which they have kept new Ph.D. graduates in apiculture out of jobs, has resulted in a real loss to beekeeping research.

The combination of government downsizing, transfers of non-bee personnel into bee jobs, and university priorities shifting jobs away from traditional apiculture has caused an alarming drop in apiculture's ability to address contemporary research problems relevant to beekeeping and crop pollination. It also has created a discouraging environment for young researchers looking for jobs working with bees.

If you are concerned about this trend, as I am, there are things you can do about it, but it's not going to be pretty. Universities and governments don't like being told what to do, even though they may be public institutions funded by your tax dollars. Universities and government laboratories will follow what they perceive as their own self-interest if left to their own devices. Your job is to make sure they feel pain when they ax bee jobs and get rewards when they maintain bee positions.

The pain is inflicted through effective lobbying. A dean is low on the totem pole compared with the head of a powerful legislative committee, and the head of a government research branch is a cockroach from the perspective of a congressman. If we are to maintain bee research, beekeepers need to find strong legislative allies at state and federal levels, and convince them that votes will come their way if they follow our advice. We have not been effective enough in this area, and that is one reason why bee jobs are disappearing in spite of the increasing need for them.

The rewards end is more positive, but also requires an organized effort by beekeepers. There is nothing universities appreciate more than money donated for research, scholarships, buildings, endowed chairs, or whatever. Simply donating money doesn't ensure that your wishes will be followed, however. You need to put appropriate strings on the money to make sure the recipient hires the type of person you want to see employed.

Try this one out on your local Entomology Department chair or state legislator: Tell him or her the beekeepers in your state will donate $100,000 toward bee research, but insist the university or government match that donation. Also insist that you have voting representation on the committee organized to disperse the funds, and on any search committee that is looking for an individual to study bees. Make certain the restrictions you put on the money are ironclad; I can think of a few burned beekeeping organizations that thought they were getting a faculty member who would do practical beekeeping research but ended up with something else entirely.

At a national level, try raising a million dollars for bee research through the Honey Board, then offer to share some of that pot with USDA bee laboratories if (1) positions at the lab are guaranteed to remain in apiculture, and (2) beekeepers have equal say with researchers in determining how the money is spent. If the government doesn't want to accept your offer, try a few universities. I know mine would bang down your door for the opportunity to get at that kind of research funding.

The decisions being made today about jobs will affect bee research for the next twenty or thirty years. We need the contributions of exciting young scientists trained outside apiculture, but we also need to maintain a solid core of personnel grounded in traditional bee management. Today, too many researchers are being hired from outside the bee area, and too few true apiculturists are getting jobs. Pay attention now, or the research infrastructure built by previous generations will soon disappear, to be replaced by scattered, overworked, and underfunded researchers unable to meet the research needs of tomorrow's beekeeping community.

PART FOUR
Life in the Research Lane

Life in the Research Lane Bee-

keeping may be a subculture in our society, but research is a foreign country with its own customs, idiom, language, ideals, and cultural assumptions. I often move between these worlds, talking with graduate students at one moment about a new finding in basic honey bee biology and with beekeepers the next about what to feed their bees. I see no inconsistency in trying to straddle the two worlds. It is much like crossing the border between the United States and Canada. At first, everything looks superficially familiar and people seem to be talking the same language. Yet, the more time you spend in the two countries, the more you realize that each is distinct and has much to offer the other.

The clearest difference between the beekeeping and scientific cultures is in the emphasis scientists place on basic research and the beekeepers' imperative to produce immediately useful management results. Bee science is a microcosm of the greater societal division between the esoteric world of the scientist and the practical demands of the taxpayers. I see this division expressed at a global level in the ongoing congressional and parliamentary debates about science funding, and at a local level as bee scientists attempt to interact with beekeepers.

This is a turf that too often degenerates into a battleground rather than a meeting place, and I tend to blame the scientists more than the beekeepers for the lack of a common understanding when beekeepers and scientists come together. At its most serious, the beekeeper-scientist boundary is maintained when bee scientists simply don't care about the immediate management concerns of beekeepers. There is an ethic of research purity that in my experience has been an undercurrent in many university departments, an ethic that pays lip service to societal imperatives while enjoying freedom to pursue curiosity-driven research wherever it may lead. The price of that underlying message when passed on to our students has been disdain for applied work, and my own feeling is that scientists are seriously deficient in our ability to perform management-oriented research that can directly benefit industries like beekeeping.

It would be an error to trivialize this message into a basic versus applied research argument, however. Basic science is and always will be the predecessor to applications, and researchers need to maintain our ability to pursue new levels at which to understand nature. Nevertheless, basic work should lead to benefits for society, and those of us who have the luxury of pursuing our curiosity need to balance that privilege with our responsibility to pay back our tax-paying funders with direct benefits. There should be a bottom line to research, and I question a research career in which that bottom line is not economically positive. Society working through governments and universities often funds a research laboratory to the extent of hundreds of thousands of dollars per year, and scientists should consider adopting an ethic to pay back that investment with usable results over the lifetime of a career.

Another area of cultural conflict between scientists and the public is the arcane world of peer review. Grants are awarded and papers are accepted for publication on the basis of scientists' review of each other, and the rules we follow usually ignore opinions outside our own culture. A beekeeper, for example, rarely if ever has the opportunity to add his or her voice to a debate about funding a particular project, and certainly does not have the chance to review a scientific paper for readability or relevance. The insular nature of scientists' review systems have led to just what you would expect: a community in which the needs of outside groups are viewed as a side issue, and the ability of outsiders to communicate with those on the inside is prevented by the dense and technical nature of our writing.

I continue to advocate cross-cultural exchanges. Why not put the public on grant and manuscript review panels and give them a real say in what gets funded and how the results get reported? If a beekeeper can't be convinced that research is relevant, even if it is basic research with a long-term potential payback, should we really be using tax dollars to fund that work? If a scientist cannot communicate results in a fashion easily understood by someone with a high school education, or is not willing at least to try, should we really be committing our tax dollars to him or her?

Another aspect of scientific culture difficult for the public to understand concerns how research is conducted and the imperative need for freedom to pursue results in an unbiased way, and with rigor. The scientific and business communities often conflict in this area, because industry wants to closely define the goals of a project while scientists want to move in whatever direction their results take them. Industry puts out contracts that are highly specific, while science needs a more flexible paradigm in which to work. While we should strive to produce a research bottom line that is positive in benefits accrued to society compared with funds invested in our laboratories, it would be a mistake to over-manage science to the point that new discoveries cannot be pursued simply because they were not prespecified in a contract.

Scientists also are as bandwagon-oriented as the rest of us. Just as we all follow the latest trends in fashion or watch the current hit show on television, scientists leap into new technologies and perspectives, following the crowd for fear of being left behind or left out. Molecular biology is a good example; the latest trend is to consider almost every aspect of biology in terms of molecules. Apiculturists' thinking about honey bees has attempted to keep pace with the broader scientific community. We no longer think of honey bees only as whole organisms, we now break them down into their individual molecules and consider using recombinant DNA technology to produce new varieties of bees. The potential to genetically engineer more useful bees has not been lost on bee researchers, but I find it comforting that the bees we use today have yet to come under the molecular knife.

A final area of cultural misunderstanding between beekeepers and bee scientists is the burden of proof each culture requires to prove a point. Beekeepers are raconteurs and tend to the anecdotal. Scientists tremble at the thought of believing an unproved statement and require replication, controlled experiments, and statistical analysis before accepting a result. A typical scene at a beekeeping meeting is a scientist presenting reams of mind-numbing data slides during a talk to conclusively demonstrate a point, followed by a beekeeper asking a question that begins something like, "Last summer, I had this hive that I got from my uncle so-and-so who lives up in the hills out beyond somewhere. He's been beekeeping for fifty years, and he said . . ." The beekeeper goes on to tell a long story about one particular hive that contradicts the years of data the scientist had painstakingly accumulated, and almost everyone in the room trusts the anecdote more than the data.

Are bee scientists and beekeepers, or scientists and any public group, from different cultures? You bet. The good news, though, is that it is not all that difficult for each of us to visit the other's country, and to understand the other's perspective. Efforts by scientists to justify our work, to include the public in funding decisions, and simply to communicate can bridge the cultural gap. A willingness by beekeepers to accept the rigor of science and the need for some scientific freedom in pursuing research would help as well. It's like most things: we can be so much more, and can accomplish so many more things, working together than any of us could hope to do separately.

I live in the very southern corner of British Columbia and sometimes drive south across the U.S. border toward Seattle. The border crossing I use is called Peace Arch because of a shared park of that name right at the boundary between the two countries, at the center of which is a large arch erected exactly at the border. Engraved on the arch is a saying that always moves me, because I grew up in the United States and have strong ties to both countries. The engraving says simply: "Children of a common mother." It refers, of course, to the fact that the governments of both countries descended from England, but I think it means much more and can tell us something about other issues besides politics.

Beekeepers and bee researchers are also children of a common mother. Both groups make their livelihoods from bees, and both have insatiable appetites for probing the depths of honey bee biology. Like Canada and the United States, beekeepers and bee researchers have much to learn from each other, and the bond of their common perspective can provide the bedrock on which to build a stronger, more interactive, and mutually supportive relationship.

23. Payback Time

ONE OF THE USUALLY PLEASANT PARTS OF MY JOB is to serve on supervisory committees that oversee the progress of graduate students. The culmination of each student's degree is a formal oral examination in which the student presents his or her research and then faces two grueling hours of difficult questions from the committee. I recently participated in one such examination for a student who had worked on bird behavior and had produced a superb thesis. I was the last committee member to ask my questions, and by the time my turn came around there really wasn't much left to ask the student about his thesis. So, I posed the following question: "I estimate that your thesis cost the taxpayers of Canada about $100,000 for your salary and research expenses. Let's say you're talking with my neighbors, who include a commercial fisherman, a retired janitor, a high school teacher, a secretary, and an ambulance attendant. Their tax dollars funded you, and they ask you to explain why it was worth it to them to have paid for this study. How would you respond?"

To my surprise, this otherwise eloquent student was floored by the question. He had answered every previous question with confidence and skill, yet he simply could not come up with even one justification for the taxpayer dollars he had received. Even worse, he seemed almost insulted that I would ask such a question, as if he and his fellow academics had some divine right to conduct research at the taxpayers' expense. Perhaps his response shouldn't have surprised me; after all, at least some of my colleagues show a similar disdain for the public trough that nurtures them, and feel that knowledge for its own sake is ample justification for the research dollars that come their way.

I thought it might be interesting to try and answer my own question: Have my job and my research for the last eighteen years been worth it to the taxpayers of Canada? I could try and avoid the question with one of the stock answers: "I've trained students for useful careers"; or "I've advanced knowledge"; or how about "I work with bees, which are important for agriculture." These answers just don't satisfy in today's deficit-ridden, over-taxed society. My neighborhood taxpayers want to know dollar values for the costs and benefits of their investment in government-funded programs. So I decided to figure out the average amount Canadian taxpayers pay me in salary and research funds every year, and then calculate the financial returns based on some monetary value for my output.

The cost to taxpayers each year was the easiest part to calculate. On average over the last five years, Canadians have funded me and my research team to the tune of $200,000 per year (all values used here are in U.S. dollars). The clear majority of those funds are for research, and pay for student and technician salaries, equipment and supplies, operating expenses for two trucks, and travel to research sites and scientific meetings. The rest is my own salary.

Basically, my job has three components: research, teaching, and administrative duties, each of which takes up roughly one-third of my time. Let's make this easy: assume the value to society of all the university committees I work on, faculty meetings I attend, forms I fill out, and position papers on university policy that I write is $0. Not a great start; one-third of my time is worthless from my neighbors' point of view!

Research and teaching, however, can be assigned some monetary value. Let's start with research. What is the annual value to taxpayers that has resulted from the research I and my students have done? I've chosen to include only three of our research projects that seemed to have the most applications for beekeeping and crop pollination: (1) queen and package bee production, (2) mass queen overwintering, and (3) pheromones. I've taken the conservative approach and said that whatever the annual commercial value generated by each project, I and my team should take only 20 percent of the credit. After all, the results of each project were implemented by beekeepers, farmers, and industry, who deserve the lion's share of the credit for taking our results and turning them into income.

Our work on package bees and queen production contributed to the bulk bee industry in Canada by demonstrating that Canadian beekeepers can rear their own queens and produce packages and nuclei for sale in the spring. Further, we developed methods to produce bees and queens economically. Queen and bee production is worth about $450,000 per year to British Columbia, so my 20 percent share of the credit produces a value to society of $90,000 per year for that research.

The queen-overwintering project investigated how to produce fall queens and keep them alive in banks through the winter. The methods developed in that study allow the individual beekeeper who employs them to earn an additional $10,714 per year above expenses, on average. If we assume that twenty beekeepers take advantage of this technology, which is a very conservative assumption, and again give me 20 percent of the credit, this project pays back $42,856 per year on the taxpayers' investment in my research team.

The pheromone work is just reaching the market. Our queen pheromone is being sold throughout North America by a Canadian company for various beekeeping uses and as an attractant to enhance crop pollination. A lowball estimate for the annual sales of these pheromones is $250,000 (my value here: $50,000), but the most significant value is in increased crop yields following spraying during bloom. The two crops for which we have been most successful at demonstrating yield increases, highbush blueberry and pears, have demonstrated average increases of approximately 5 percent following pheromone spraying, producing an average profit increase of about $1,000 per hectare. If I use a hyperconservative estimate and say that one thousand hectares in North America show yield increases following pheromone sprays, I calculate a profit increase of $1 million, of which I can take credit for $200,000 per year that our work has contributed back to the taxpayers.

I also teach, and it took me some time to decide how to put a dollar value on my teaching. Then it hit me: my courses are worth what students are willing to pay to enroll in them. Total tuition for a

Simon Fraser University course is $157.50 (yes, tuition in Canada is a bit lower than for U.S. universities, which actually works against me in these calculations!). I also teach courses such as Bee Masters, a week-long course in advanced beekeeping that we offer every other February, which costs about $450 for the week, including tuition and room and board. Grand total: $30,937 paid by students to take my courses.

Another component of my job is delivering lectures to beekeeping groups, usually in Canada or the United States but sometimes as far away as New Zealand or Australia. My "value" for these lectures can be calculated as the amount beekeepers are willing to pay for me to attend their meetings. I attend about nine beekeeping meetings a year, and my average airfare and accommodation expenses total about $4,500 annually, which seems to be a realistic monetary value for this service.

Adding it all up, I calculate a value of $418,293 return per year to the taxpayers for their $200,000 annual investment in my salary and research expenses. I think these calculations are extremely conservative, but even if I reduce my value by one-half of what I calculated ($209,146), or dismiss the largest single component of these calculations—the $200,000 value for pheromone-based yield increases for blueberries and pears (leaving $218,293 in returned value)—the taxpayers' support for my salary and research program still leaves society in the black.

University administrations and individual faculty members have resisted these types of calculations, partly out of a philosophical belief that our benefit to society is self-evident, and partly out of fear that we might come out on the taking rather than the contributing end. The luxury of receiving copious amounts of money from taxpayers for unjustified expenditures is no longer available, however. Governments running huge deficits are trying to "rationalize" programs and are demanding that each funded unit provide some payback to the funders, much as stockholders in private business demand dividends and rising stock prices from their investments.

I'm not sure this approach is bad, partly because I believe we all should be able to justify the money we're paid by our bosses, whether they be the stockholders of our company or the taxpayers on the streets where we live. But I also think universities are selling themselves short by not aggressively promoting the value of the education they provide. I don't know about your local university, but the one I work at provides real returns to the taxpayers in terms of new products, jobs, and consulting services that enhance local businesses and governments.

Taxpayers, don't be shy about demanding some value for your investment in your universities. Professors, don't be reluctant to put a dollar value on your work. Education does have an important role in society that goes beyond the financial bottom line, but nevertheless a strict accounting of that bottom line should be a positive one for most of us who teach and do research. We return good value to the taxpayers who support us, and I think we have an obligation to our funders and to ourselves to convince the taxpayers that their investment has been worthwhile.

24. The Bottom Line

I'M FINALLY GETTING OLD ENOUGH TO BE NOSTAL-gic, a state of mind in which you remember how everything was better in the good old days. I remember when country music was something you listened to in your car late at night, driving the back roads. Today, I sit on my couch watching television country music in video format, with strange images on the tube riveting my attention and putting the music in the background. I remember receiving a letter and taking the time to read it, think about it, and eventually write a response. Today, if I don't respond within minutes to the whine of the fax machine and the beep of my electronic mail system, my "correspondents" assume that I'm sick, retired, or dead.

And I remember when beekeeping was a simple occupation, when a good queen and sound management practices could provide an income. Today . . . hmm, maybe things haven't changed so much after all.

The bottom line for commercial beekeepers is still making money, and the bottom line for bee researchers remains providing information to help them make that money. That aspect of things hasn't changed at all. The world we live in has changed; though; it has become more complicated than it used to be. Government regulations, new diseases, paperwork and taxes, urbanization, and big-business farming have created a more complex environment in which to manage bees; however, the solutions to these problems may not be as complicated as they appear. Good bee management decisions, in colonies with a good queen, can still earn a beekeeper's living in the 1990s. Bee research has changed, too. Bee scientists have gotten away from what I call the Bottom Line approach, management-oriented research that can tell a beekeeper which management system will earn more dollars at the end of the season.

I'm not writing this essay to criticize my fellow researchers, or to bemoan the sad state of bee research today. I think bee research is healthier now than it has ever been. I'm amazed at the training and overall quality of our young researchers, and the scientific tools they have to work with dwarf anything that was available in the past. Further, the problems the beekeeping industry faces can and have benefited greatly from contemporary research contributions. But the fast-paced world has influenced researchers to forget to finish what they start. We make, publish, and then abandon new discoveries without ever asking the Bottom Line question: Will a beekeeper make more money at the end of the season because of my research?

I searched for recent examples of Bottom Line research to illustrate good management studies, but I couldn't find many examples. So I went back to the olden days, the 1980s, and there found some real chestnuts, excellent examples of how old-fashioned management research made important contributions to beekeepers' livelihood.

One of my favorite Bottom Line studies was conducted by a former student of mine, Cynthia Scott-Dupree, in the scenic Okanagan

Valley region of British Columbia. Cynthia was asked by the Okanagan Valley Pollinators Association to answer a very simple question: Do we make money moving our bees to pollinate orchard crops in the spring, or would we make more by leaving the colonies in their yards?

She compared several different management systems, including (1) moving bees for pollination; (2) honey production alone, with no pollination moves; and (3) package bee production from pollination units, maintaining colonies at or above minimum pollination strength following package shaking. Cynthia examined labor, feed, moving, and equipment costs for each system, as well as income from honey production, pollination contracts, and package bee sales. In addition, she measured biological characteristics such as brood area and colony population at the end of the season to determine whether any of the management schemes resulted in poorer quality colonies. The results were fascinating: The most intensively managed bees—those used for honey production, pollination, and package bee production—yielded the most income, with no real loss in colony vigor at the end of the season. In fact, the intensively managed colonies were in *better* condition than those used only for honey production! The Bottom Line: Honey production alone yielded an average gross income of $12 per colony; honey plus pollination yielded $67 per colony; and honey, pollination, and package bee production yielded $87 per colony. These results were taken to heart by the Okanagan Valley beekeepers, many of whom not only continued to pollinate, but now also produce packages and nuclei each spring.

Good Bottom Line research was also conducted by Don Mac-Donald, the former Alberta provincial apiculturist, and George Monner, an economist, in the Peace River region of Alberta. This area is far to the north, and the combination of long summer days and good bee forage produces the highest colony yields in the world, often surpassing three hundred pounds per colony. Beekeepers in the Peace River area traditionally purchased package bees from the southern United States each spring, harvested all the honey at the end of the summer, and then killed the colonies. The impending threats of mites and Africanized bees inspired MacDonald and

Monner to investigate whether this package bee system was providing the best income, or whether overwintering systems might yield equal or better returns.

Their study involved a detailed economic analysis of a two-thousand-hive operation, and included such items as depreciation, precise costs, labor, interest on bank loans, and, of course honey production. Overwintering required considerably more long-term capital investment than package bee management because it was necessary to build and operate an overwintering building. However, the package bee operation had higher short-term costs to purchase packages each spring. The Bottom Line: The overwintering system provided a return over costs of $24.70 per hive versus $13.75 for packages. This study gave our beekeepers considerably more confidence in switching from package bees to wintering, and its methods are still used by beekeepers to calculate the economics of various wintering systems.

Another good example of Bottom Line management research also involved wintering, and was conducted by Basil Furgala and his students Mark Sugden and Steven Duff in Minnesota. They evaluated colonies wintered outdoors using two different hive sizes (two- versus three-deep hive bodies) and two commonly used winter packing methods, insulite boards versus cardboard winter cartons. They evaluated winter survival, spring build-up, and honey production. Further, they recognized that apiary site and seasonal differences affect colony performance, so they conducted their study at two different locations and for three seasons to make sure they had fully evaluated the impact of their treatments.

Furgala and his students calculated a mean productivity index for each treatment that took into account the number of colonies surviving each winter and the subsequent honey production from each colony. They found no differences in colony productivity due to the winter packings they tested, but the three-high colonies yielded approximately five-hundred pounds more honey per colony than the two-high units over the three-year study period. If we assume the 1996 average price of $0.89 per pound for bulk honey sales, an operation of one thousand colonies using the three-high system would have made $445,000 more over the three-year period ($148,000 per year) than the two-high. Thus, a very simple management maneuver, threes versus twos, made an enormous difference in the Bottom Line.

Furgala's laboratory did another management-oriented study that I have always admired because I've never had the nerve to do it myself. They evaluated six honey bee queen stocks used in Minnesota by purchasing queens of different lines from commercial queen rearers and introducing them into experimental colonies. This type of study obviously takes courage, because in the end the queen rearers whose stocks produce the most honey will love you, and those at the bottom will never speak to you again!

They examined a number of stock characteristics, including aggressiveness, tendency to swarm, queen loss, and honey production. Again, they did it right by managing everything the same way and conducting the study at several locations over a two-year period. Without mentioning the name of their stocks and producers (I'm not nearly as brave as Furgala!), I will say that they demonstrated that some of the stocks tested were not suitable for Minnesota. The best queens headed colonies that produced annual honey crops of eight to thirty pounds per colony more than the other queen stocks. For a thousand-hive operation, again assuming honey at $0.89 per pound, the better queen stocks would have yielded $7,120–$26,700 more income each year. The use of the best queen stocks would have provided the difference between survival or bankruptcy for a beekeeper operating an economically marginal business, or the difference between a winter vacation for the family in Hawaii versus Duluth, Minnesota, for the beekeeper in the more profitable situation.

The Bottom Line: There's still a place for economically oriented management research in the beekeeping community. North American beekeeping is enormously diverse, and there are many ways to make money with bees—and to lose it. Bottom Line management research can make the difference between the making and the losing, and is a deceptively challenging art form for scientists searching for ways to be intellectually stimulated while conducting useful research. Indeed, there is a real elegance to a good management study, a satisfying sense that simple solutions for complex problems are still out there. Maybe it's my simple mind, but I'm glad there are still radio stations playing late-night country music, letters to write, and Bottom Line research to be performed. So much for nostalgia: it's all still out there, just waiting for us to do it.

25. Peer Review

BELIEVE IT OR NOT, SCIENTISTS ARE PEOPLE, WITH our own set of rules by which we practice our craft. Scientific practice is highly personalized, and although scientists like to consider ourselves rational and data oriented, our individual quirks and biases influence our work more than we care to admit. In that sense we are artists using the medium of scientific publications rather than canvas or clay to express ourselves. Indeed, the authorship of the publications we produce can be easily identified by signature properties of style, form, subject and perspective, much as a Rembrandt, van Gogh, or Monet is recognizable by even the most untrained viewer. For example, regular readers of my *Bee Culture* columns can easily tell whether an article is mine by both its style and its perspective. Science and art perform different functions, however; science is supposed to provide unbiased and fact-oriented analyses for society to use to advance knowledge, promote economic advancement, and make decisions about how to manage the world.

This conflict between what society thinks science should do and be and the reality of science as art is resolved through the process of peer review. When a scientific publication is submitted to a journal, the editor routinely sends it to two or three scientists working in related fields for review. The identities of the reviewers are kept secret from the authors of the manuscript, although the reviewers' comments and recommendations are transmitted anonymously to the authors. Generally, an article is accepted for publication only if the reviews indicate that it contains new information, the work was conducted properly, the data analyzed correctly, and the interpretation of the results fit the data. The process of peer review also extends to grant applications, which can be sent to ten or more reviewers. Thus, the process of peer review is an essential component of scientific practice, since it mediates who gets funding and who gets to publish.

What does peer review protect scientists from? For one thing, it differentiates between perspective and bias. Research is very personal—and should be—but a valid perspective can easily degenerate into the "ax to grind" syndrome, in which two research camps degenerate into advocacy research, with each side designing projects that prove in subtle ways that they are right and the other side is wrong. The bee research world has had its share of quarrels in which personalities have elevated healthy differences in perspective into pitched battles between "true believers." Many a career has foundered on the rocks of these controversies, but in the end, peer review of grants and papers is an effective mediator. Continued ax-grinding research that is not backed up by solid data rapidly leads to loss of grants and rejection of publications, so peer review serves as a regulator to separate good science from advocacy research.

Peer review also encourages research to be conducted by independent laboratories, thereby protecting society from industry-funded, product-driven research that attempts to prove a product "works." Some of you may grouse about your tax dollars going to support government or university research laboratories, but would you rather rely on industry research to tell you whether menthol and fluvalinate are effective or on researchers with no economic investment in the product? A good example of this credibility issue is the label recommendation that Apistan strips be discarded after a single use. Many beekeepers reuse Apistan strips, believing they are being misled into purchasing new strips when the old ones are perfectly good for numerous applications. Recent independent research conducted by a Washington State government laboratory showed that the recommendation to use strips only once is correct; Apistan strips lose much of their potency after a single application. This research is credible because it comes from a source with no ax to grind.

Peer review has another function that is underutilized: it protects all of us from bad writing. I am a highly trained, educated, and experienced scientist, yet I can't understand many of the scientific papers I see in technical journals simply because they are poorly written. I figure that if even I can't make heads nor tails of these articles, how is the average beekeeper going to figure them out? The common failure of scientists to communicate effectively is largely the fault of poor training in the communication arts, but peer reviewers nevertheless have a responsibility to reject manuscripts that are not clearly written. I have occasionally returned manuscripts to editors without review, asking that the editors insist on clear writing before asking me to use my time to review a paper. Perhaps journal readers should do the same; next time you read a bee article that is hard to understand, return the issue to the editor and ask that it be rewritten so you can understand it!

Peer reviewing itself is an art form. It requires considerable tact and a willingness to suspend your opinions and provide an unbiased review of a competitor's work. Just as authors of manuscripts can have an ax to grind, so can reviewers. Oddly enough, however, the peer review system works well because of the unwritten but heavily and informally policed ethic that reviewers should not permit their own perspectives to degenerate into biased reviews. Yes, there are occasional miscarriages of reviewing justice. When that happens, editors are very willing to have a manuscript re-reviewed by additional reviewers to determine whether or not the review was fair. Editors and granting agencies remove the names of reviewers from their lists if there is repeated evidence of biased reviews; a few angry letters from authors, if substantive, are usually enough to remove a problem reviewer from the process.

I do have one radical suggestion that I think would substantially improve the peer review process: I believe beekeepers should serve as voting members of granting agency review panels for bee research, and also that they should play a role in reviewing manuscripts for beekeeping journals such as the *American Bee Journal*, *Journal of Apicultural Research*, and *Apidologie*. That's right; you read it correctly: let's take some of the peer review process out of the hands of scientists and share it with interested outsiders.

My opinion on this subject was forged through meetings I participated in to allocate $300,000 in research funds provided by the Canadian government to the Canadian Honey Council. The council wisely recognized that it needed expert help to evaluate the numerous grant applications that came in, but equally wisely insisted on having strong influence in determining what got funded. A joint committee made up of researchers, beekeepers, and extension personnel was formed to evaluate the proposals, and I was impressed with the quality and insight of industry representatives during the evaluation process.

Beekeepers have a real stake in bee research, and can also contribute valuable input into grant and manuscript evaluation. While there may be some technical information that requires scientific training to understand, I would hesitate to fund any bee research grant that could not be clearly explained at some level to a beekeeper. You can bet that if the grant proposed is incomprehensible, the published results will be no clearer—and thus will be worthless to beekeepers. Is there something wrong with taxpayers having some control over how their tax dollars are spent? Why shouldn't citizens have a vote in granting agencies such as the U.S. Department of Agriculture, the National Science Foundation, and state organizations? Such a system would select for those who can explain and justify their research, and I don't think there's anything wrong with that.

Similarly, why shouldn't beekeepers have a role in deciding which manuscripts deserve publication in cases where the research was funded by their taxes or contributions? Perhaps the scientific evaluations are best done by scientific peers, but shouldn't there be a clarity component in manuscript evaluation that screens for readability by interested commodity and public groups? There may be valid reasons for publishing research in language that is too technical for the layman, but shouldn't the public insist on companion summaries that anyone can understand before a journal agrees to publish a manuscript?

With this in mind, I propose the following changes to the peer review system: (1) All granting agencies should have nonscientists making up 25 percent of the voting members on decision-making committees. Grants for applied bee research should be reviewed by beekeepers, and more basic grants such as those funded by the National Science Foundation should be reviewed by citizens from a broader range of professions. (2) Journals publishing research with beekeeping applications should include one beekeeper in the review process for every manuscript, and should insist on clear writing in addition to scientific merit as a component of manuscript review. (3) Granting agencies should withhold 10 percent of all grant funds until the project is completed and a beekeeper-friendly and easily understood report of the work has been prepared and either submitted for publication or otherwise distributed to the beekeeping community.

Peer review has served the scientific community well over the years in allowing scientists to express their own styles while maintaining scientific integrity. The reviewing process can do a better job serving beekeepers, however, and I hope some of you editors and grantors out there will give the beekeepers an opportunity to participate in the process, and broaden the utility of this important aspect of the research profession.

26. Behavioral Ecology

SCIENCE MOVES FORWARD IN FITS AND STARTS; PEriods of stagnation are followed by bursts of discovery. This creates the bandwagon effect, in which a new approach is eagerly embraced by the up-and-coming generation of scientists. Scientific fads come along regularly, and either die out if they prove to lack substance or grow and prosper if they provide new knowledge and technical advances.

The field of behavioral ecology is one such fad that has stayed around. It began in the 1970s, at a time when many biologists were becoming frustrated with the separate fields of behavior and ecology. Animal behavior had become too laboratory oriented and "psychological" for the young rebels, who wanted to study behavior in the field. Ecology had become isolated from the behavior of animals and plants in the real world, focusing on community structure and species diversity more than on interactive questions about behavior and evolution under different ecological regimes. The field of behavioral ecology was born to bridge this gap by studying how animals behave in nature.

Behavioral ecology is on my mind because of a comment made by my *Bee Culture* editor, Kim Flottum, in his March 1995 "Inner Cover" column. He wrote about an article concerning how bees regulate pollen foraging that was published in *Behavioral Ecology and Sociobiology*, the flagship journal of this field. Kim was interested in this article because he thought it might provide information useful for the management of commercial pollen collection, but he was disappointed because, as he put it, "This is the type of research a scientist conducted for his or her own gain as far as I'm concerned."

The subject of relevance in scientific research is always on my mind, but Kim's comment especially caught my notice because my students and I often publish our more basic research in *Behavioral Ecology and Sociobiology*. In fact, we published an article on pollen regulation in that journal in 1992, although I have no idea if it was the article Kim was referring to. You'll understand why Kim's remark caught my attention if you read my columns on a regular basis, because I often express my concern about how scientists rarely discuss the applications of our research in scientific journals.

Just for fun, I dug out our 1992 pollen-foraging article. This research was conducted and the article written by Jennifer Fewell, who was at that time a postdoctoral fellow in my laboratory and is now an assistant professor at Arizona State University. Right away I could see how Kim or any beekeeper who read this article might have concerns about relevance. The very first line of the summary says, "To place social insect foraging behavior within an evolutionary context, it is necessary to establish relationships between individual foraging decisions and parameters influencing colony fitness." It would take a real stretch of the imagination to believe that a beekeeper reading that introductory line would find relevance for beekeeping in it, or even that he or she would continue to read on to the next line.

Nevertheless, this article had some significance for bee management and did provide basic information that should lead to improved income for beekeepers. I admit that a beekeeper who read the entire article might not have seen dollar signs jumping out from the page, but that was because we didn't write the article for beekeepers. Rather, it was written in the style and using the jargon of behavioral ecology, a dense thicket of buzzwords, assumptions, and shared paradigms that is easily traversed by those who work in the field but is almost impenetrable by outsiders.

The research we reported in this study used two groups of colonies that were identical in every way except for the amount of stored pollen they contained. We equalized brood areas, adult populations, honey, and so on, but manipulated colonies so that one group contained a lot of stored pollen (about three full frames) while the other group had only a little stored pollen (about one-eighth of a frame). Then we examined foraging from colonies in the high and low groups, paying particular attention to total numbers of foragers, proportion of bees foraging for pollen, and pollen load size.

The results were clear and dramatic. While the total number of foragers was the same in the high- and low-pollen colonies, the proportion of foragers returning with pollen was much higher in the colonies that had low pollen stores. In addition, the workers from the low-pollen colonies returned with larger pollen loads than those from colonies with larger pollen stores. They also made shorter trips and spent less time in the colony between trips. All these data indicated that foragers from colonies with little stored pollen put much more effort into pollen foraging than workers from colonies with a lot of stored pollen.

How you interpret these results depends on whether you're a behavioral ecologist or a beekeeper. We published this article in a behavioral ecology journal, so our discussion of the results focused on questions of interest to scientists in that field. We treated pollen as a currency, and discussed the results in terms of strategies and fitness. For example, the last part of the article states that "despite its importance as a food commodity, pollen has traditionally been overlooked in foraging models addressing fitness questions. Our findings that 1) pollen foraging activity is tightly regulated, and 2) pollen stores have a measurable relationship with fitness variables, suggest that models of the evolution of foraging in social insects can be greatly enhanced by consideration of non-caloric foraging rewards such as nitrogen."

This dense, jargon-ridden ending masks an important implication of this study for beekeeping. Put more clearly, we found that workers from colonies from which pollen had been removed collected more pollen. Such colonies would be better pollination units because workers foraging for pollen are more effective at pollinating most crops than workers foraging primarily for nectar. Thus, beekeepers could remove some of the colony's pollen store before moving bees to a crop for pollination, and charge higher pollination fees for these superior units.

I'm sure that last line about higher fees caught your attention, beekeepers, where academic phrases about fitness left you cold. Same study put into a management context, and suddenly you're interested. Indeed, I talk about this study often to beekeeping groups because rental fees for pollinating colonies are set much too low. The kind of manipulation we discussed in our *Behavioral Ecology and Sociobiology* article provides a method by which beekeepers can add value to pollination units, and thus increase their income. A beekeeper performing this manipulation should be charging at least ten to fifteen dollars more for that unit, and even at that the grower is getting a good deal. The only extra cost to the beekeeper is the five or ten minutes it takes to remove a few pollen frames from each colony, which provides an excellent financial return for the labor.

Our article is not the only article on pollen foraging published in *Behavioral Ecology and Sociobiology* that could be criticized for its apparent lack of relevance to beekeeping. For example, Scott Camazine, formerly a graduate student at Cornell University and now an assistant professor at Pennsylvania State University, published an article in 1993 demonstrating that worker bees do not respond directly to the level of the colony's pollen reserve. Rather, workers likely perceive stored pollen levels through interactions with nurse bees in which food is exchanged. Camazine's article is similar to ours in being pitched to the behavioral ecologist audience, and is tough slogging for beekeepers. However, his article is similar to ours as well in having interesting potential applications. For example, if we could determine the chemical cues being passed between nurse bees and potential pollen foragers, we would be able to use synthetic versions of bee-produced substances to stimulate pollen collection, again increasing the pollination fee that could be charged for those colonies.

Thus, Kim's complaint about the journal *Behavioral Ecology and Sociobiology* has little to do with the substance of the research reported and a lot to do with how it was presented. Scientists can speak in two tongues: the obscure jargon of basic science and the more easily understood and directly relevant style of applications. The work we do often has both levels of interest, the fundamental and the applied, but we keep them separate in our writing. Quite frankly, university faculty prefer to publish in the academic journals because we get promotions and tenure for *Behavioral Ecology and Sociobiology* articles, and no recognition at all from our colleagues, chairs of our departments, and deans of our faculties for a *Bee Culture* article.

I staunchly defend Fewell's and Camazine's papers. Although they are fundamental science, they have interesting applications and commercial relevance. However, I question our academic system, which pushes fine young scientists to publish their work in a way that only other academics can understand, and ignores or even discredits the flip side of basic research, the management implications.

Perhaps we should start a new journal in this field and call it *Behavioral Ecology and **Management***. Manuscripts submitted to this journal would have to include the basic perspective *and* the applications, and be written in a jargon-free style understandable by anyone. Reviewers for this new journal would include scientists, representatives of relevant industries, and maybe even people randomly chosen from the phone book. Finally, publications in this new journal would become the most highly prized citation in a faculty member's curriculum vitae and lead to rapid promotion and tenure.

Maybe it's time for a new bandwagon: the marriage of basic science and applied science, and mutual respect between the disciplines. This new fad would have simple, clear writing as its trademark, and broad relevance as its only mandate. And to start it off, I suggest the two editors of *Bee Culture* and *Behavioral Ecology and Sociobiology* hold a summit meeting and agree to publish a joint trial issue together. Who knows; it could be the start of something wonderful.

27. Things I'll Never See

IF AN AWARD WERE TO BE GIVEN OUT FOR THE PERson of the Century, it would have to go to a generic Scientist. This has been the century of science and technology. Simply unbelievable progress has been made in the human condition as a result of advances made through scientific research and the application of technology derived from that research. Just think for a minute about the profound difference between our lives and the lives of people a century ago. We routinely drive cars, fly in the air, and can travel to the moon, although at some expense. I am a middle-aged person, yet I have been to very few funerals because modern medicine saves most of my cohort from dying before we reach our seventies and eighties. The manuscript for this book can be sent to Cornell University Press via the electronic highway in seconds; only ten years ago it would have taken days or weeks to get there by mail.

Many of our ancestors' most time-consuming tasks can be disposed of today with only the effort it takes to push a button: the heat comes on without our chopping wood, the dishes get washed by a machine while we watch television, even our waste gets whisked away at the gentle push of a handle.

I was thinking about this last summer while driving with some of my students out to one of our apiaries. The conversation was rambling in all sorts of directions, as it tends to do when you spend too much time in the cab of a pickup truck, and eventually we hit on the subject of science and bees. I began to wonder whether there were any limits to what science could do for beekeeping. In light of the astounding inventions of the twentieth century, it is hard to imagine an area where science has any limits, given enough time and human ingenuity. Luckily, I have this little inner voice that distrusts anyone, including myself, who claims they can totally solve a problem by applying science to it. This voice went public that day, and out of the blue I said to my students, "There are three things I'll never see: a queen finder, a better queen than we have today, and any major improvements in beekeeping equipment."

The first of these, a way of finding the queen in a hive, would elevate its inventor to the status of a Langstroth, the inventor of the modern bee hive, in the beekeeping world. Unfortunately, I don't believe there is a way to find, attract, or trap the queen except for the old-fashioned, nontechnological way: open the hive, look through it frame by frame until you find the queen, then pick her up with your fingers. Yes, you can make the job easier by restricting the queen to one hive body using queen excluders, and yes, you do develop an intuition about which frames to look on for the queen, but no one has invented, or ever will, a practical, high-tech way to locate the queen.

It certainly won't be for lack of interest or trying. I have seen or heard proposed all sorts of wild devices, some of which show considerable inventiveness and ingenuity. One idea that I invariably hear at beekeeper meetings is to use synthetic pheromones to attract and trap the resident queen in a colony. This idea sounds intriguing, particularly since worker bees can easily be manipulated with pheromones to go almost anywhere you want to send them. However, the concept of attracting a queen to a within-colony trap has one major fault: queens are not naturally or predictably attracted to any phero-mone within a colony. The most we can say about queen attraction to pheromone is that a resident queen may recognize the presence of a foreign queen by her odor. Yet, how many times have you seen two queens walk right past each other in a hive and totally ignore each other's presence? Queens may orient outside the nest to the worker pheromones given off as a swarm cluster forms, but they are not re-sponsive to these odors within the context of a hive. Even should some highly attractive odor be found, and a trap designed to capture an attracted queen, it would have to be queen-specific to prevent workers from rapidly filling up the trap.

Other, more mechanical solutions to queen finding have been proposed and even constructed, but they all suffer from overdesign by being too expensive or cumbersome to use on a routine basis. For example, a beekeeper once showed me a wheelbarrow modified into a shaker box. An entire colony could be placed in the box and shaken through a queen excluder, leaving the queen behind. It worked only some of the time, took about ten times as long as going through the hive by hand, and disrupted the colony for the rest of the day, but otherwise it was a great device! Beekeepers also dream about a magic glue-on label for the queen that could be recognized by a scanner, which could pinpoint the queen's location on a hive matrix projected onto a computer screen. Sure, it probably could be done, but think of the expense involved, and be sure to include the high probability that such high-technology equipment would break down under real-world beekeeping conditions. No, I don't think I'll ever see a queen finder that works better than the human eye coupled with a hunch as to where she might be found.

I also don't think we'll ever see a better queen than those used commercially throughout the world today. This statement is not meant to diminish the importance of good queen selection, breed-ing and rearing by commercial queen rearers; maintaining the qual-ity of queen stock used for beekeeping will always be important. However, media reporting about science has left the impression that modern molecular biology can take genes from various organisms, mix them up in a test tube, and produce new superorganisms that will solve all the world's medical, environmental, and agricultural problems. We tend to forget that there has been a natural method of mixing genes operating for a few billion years; it's called sex. Queen breeders and scientists have already influenced our queen stock through controlled open mating and instrumental insemination, and we have greatly improved queen quality over the years by controlling which drone inseminates which queen. But I question whether we can produce queens superior to those we have today.

The limitation here is that a colony is a very complex system, and selecting for one or a few beneficial characteristics may be detrimental for the overall colony functioning. Disease resistance is a good example. Have you ever wondered why the hygienic behavior found in bees resistant to American foulbrood has not become a standard component of commercial queens? Part of the answer is that these hygienic colonies have other characteristics that make them undesirable for beekeeping, such as being overly aggressive and producing relatively little honey. Today's search for tracheal mite– and *Varroa*-resistant bees may lead to somewhat resistant colonies, which would be useful, but will those colonies be any better in the end than the best in contemporary beekeeping? I doubt it. Continued queen selection may produce colonies with characteristics that might be desirable for future beekeeping situations, but I would be very surprised if the new queens produce more honey at less cost than the queens heading today's colonies. Different, maybe; better: I don't think so.

Finally, I don't think we'll ever see another major advance in beekeeping equipment with the significance and impact of the Langstroth hive or radial extracting equipment. Beekeepers certainly will make minor modifications to these and other pieces of equipment that will make them easier to use, but can you imagine a piece of equipment or new hive design that would improve on the basic concepts used in modern beekeeping equipment? The Langstroth hive is so soundly based on natural colony design that I can't come up with even a highly speculative alternative to that structure. For extracting, can any of you envision a better way of getting honey out of the comb than uncapping it and spinning the frame?

Nevertheless, the best part of this pickup truck conversation with my students was that they disagreed with me. Although they couldn't come up with specific ideas that would find queens, produce better queens, or result in a new equipment design, they argued vociferously that major advances in these areas are still possible. After all, they said, someone with my attitude a century ago would have denied the possibility of computers, pickup trucks, antibiotics, and migratory beekeeping, yet all of these are found in today's beekeeping world. I wish I could be around at the end of the next century as some beekeeping historian is writing one of those "beekeeping a hundred years ago" columns. I hope that historian finds these predictions about things I'll never see and writes a humorous article about just how wrong I was. It would greatly please whatever is left of my bones if the things I said I'll never see become routine management tools for future generations of beekeepers. I would be especially thrilled if one of those students in the pickup truck, or their academic descendants, was the one to make me eat my words.

28. Recombined Bees

POLYMERASE CHAIN REACTION. RECOMBINANT DNA. Genetic engineering. These fancy terms describe new techniques in biology that are changing the face of science. Behind these confusing terms is a simple concept: Current technology allows us to take the genetic material from one organism and place it into another, without using the old-fashioned method, sex. This is easiest to do with simple organisms such as bacteria, but we also can take a piece of frog DNA and put it into a mouse, or a piece of human DNA and put it into a rat, or even a piece of bee DNA and put it anywhere we want. The implications are fascinating: a mouse that jumps like a frog, a rat that thinks like a person, and maybe a fly that collects honey are all just waiting to be created for humanity's benefit.

Media stories about the rapid progress of molecular biology would have us believe that the wildest of combinations are possible, and suggest that the only limits to manipulating life itself are the restrictions put on scientists by a concerned public, or perhaps more important, the need for funds to conduct some of these wacky gene transmigrations.

In theory, recombined bees valuable for beekeeping could be created. Wouldn't you purchase a queen that produced antibiotic-secreting workers, created by taking bacterial genes that produce antibiotics and introducing them into a bee's genes? Even better, how about taking the genes that control the ability of the Asian honey bee *Apis cerana*, to groom *Varroa* off of bees and out of the colony, and inserting them into our own *Apis mellifera*?

The possibilities are endless, and the techniques to accomplish these gene recombinations are commonly practiced by contemporary scientists. Let's say we want to create a bee that produces its own antibiotic to control American foulbrood. First, we find a bacteria that naturally produces our desired antibiotic. Then we make a map of the gene that controls antibiotic production in the bacteria, and follow this genetic map to cut out the gene that contains the antibiotic production code. Then we find a vector to carry this gene into the bee, such as an otherwise harmless virus. We attach the antibiotic code to the virus, and then infect the target bee with the virus. When the virus enters a cell of our target queen, it carries the bacterial gene to the bee's gene, where the antibiotic code can merge, or recombine, with the bee's genetic material. Voila! We now have a bee that carries its own machinery to produce an antibiotic, and American foul brood is history!

But hold on; is it really so easy, and is it wise? Molecular biologists sometimes remind me of a tricky mechanic I used to go to. I know as much about car repairs as most beekeepers know about molecular biology, and terms like *carburetor, piston, spark plug*, and *differential* mean as much to me as molecular biology terms like *plasmid* and *polymerase chain reaction* mean to the average Joe. This mechanic would snow me with terms, coming at me hard and fast until I was convinced my car was near death. The repairs always took longer than he promised, cost double or triple his estimate, and never seemed to work; I invariably had to return for even more expensive repairs to fix the repairs that hadn't solved the problem in the first place.

Molecular biology can be like that. Gene jockeys will snow you for hours about the great technology at their fingertips, the speed at which they can create a recombined organism, the low cost, and the incredible benefits that will result from their artistry. It doesn't work that way in practice, however. Yes, molecular biology has produced important scientific progress, and yes, it has become a tool in medicine and agriculture, but no, Superbee is not at our fingertips, for many reasons.

First of all, the technology is not as simple as molecular biologists would lead us to believe. It requires incredible precision, and a great amount of luck, to find the right gene, cut it properly, and transfer it first to a virus vector, and then into a bee. It's not hard to do this from one bacteria to another, but larger organisms such as bees are enormously more complex than bacteria.

Second, most characteristics, such as antibiotic production and grooming behavior, are controlled by many genes, and they all have to be transferred to get the desired result. It's even more complicated because genes involved in a particular trait are generally found in different places, and somehow they all have to be removed and transferred together.

Third, an artificially recombined bee would not necessarily be better than a natural one. Take disease resistance, for example. We already have bees that are resistant to such diseases as American foul brood, but they have not become popular among beekeepers because they don't produce honey very well and are fairly aggressive. The same problems could result from an artificially engineered bee. We might impart disease resistance by transferring genes from one organism to another, but such resistance might come at a high cost to the rest of the bee's activities. Let's take our antibiotic-producing recombined bee again. It would be quite expensive for a bee's metabolism to produce antibiotics, and our engineered bee might well be resistant to AFB but not have enough energy left over to rear brood, build comb, and forage for honey.

Fourth, and perhaps most important, the artificial creation of recombined organisms is very, very expensive. Cost-benefit: hope for an enormous benefit, because the cost of producing recombined organisms is way out there in outer space. Molecular biology laboratories are filled with high-priced equipment that requires hordes of technicians to operate. The slightest mistake and it's back to the beginning, and mistakes are very easy to make in this high-technology field. The production of even a simple bacteria that produces an antibiotic might cost tens of millions of dollars, and getting that bacteria into a bee might take ten times the total amount of money put into bee research in a decade. It's just like an estimate to fix your car or renovate your house: triple the quoted price and the time to completion, and you're getting close to reality. Beekeeping is just too small an industry to justify the enormous expense involved in creating a recombined bee.

Molecular biology still has a long way to go before I'll be convinced that we're going to see a genetically engineered bee. Nevertheless, the field does have some significance for bee management. Molecular biologists haven't given us a super recombined bee, but they have made very significant contributions to beekeeping, and will probably make more as the technology improves. One good example is in certification of bees as non-Africanized. In the future the major barrier keeping U.S. beekeepers from exporting bees to Canada is going to be Africanized bees. Canadian beekeepers are reluctant to import queens or packages if there is even a slim possibility that these might be Africanized. Further, the current procedures to certify bees as European are not very reliable. Enter molecular biology: Africanized, European, and hybrid bees have characteristic gene patterns that can be readily distinguished by molecular biology techniques, and the methodology can be made simple and automated enough to provide an accurate system to certify bees as non-Africanized.

The cost-benefit ratio here is much more favorable than for our recombined bee. The potential market for queens in Canada alone could easily reach three to five million dollars a year, but the development expenses to complete an assembly-line type of molecular certification procedure should cost only a few hundred thousand dollars (triple my estimate, but it's still worthwhile). Even if Canada required every queen to be certified, and the test cost fifty cents per bee, it would still be economically feasible. Also, certification procedures for moving bees within the United States would benefit from a more accurate test. There's money to be made here by a private company conducting certifications using molecular means, and by beekeepers selling accurately certified queens to new markets.

The problem with molecular biology is not its real or potential utility, but rather its perspective. The situation with bees is just a microcosm of a much larger issue, the seductive nature of technology. Scientists are just like the rest of us; they leap head-first into the newest, latest, and "sexiest" methods, sometimes without thinking about whether new technology is warranted. We tend to lose sight of the fact that technology is a tool. Sometimes the newest tools are very useful, but not when they are used largely because they are novel rather than because they are the best solution to a problem.

The honey bee examples I gave above provide some guidelines as to where molecular biology can fit into beekeeping and where its use is overkill—and perhaps even damaging. Molecular techniques are excellent tools for identifying types of bees for study purposes, and for use in certification programs. They are also highly valuable for helping us understand how bees function at genetic, physiological, and even organismal levels. However, old-fashioned bee breeding is still a much more cost-effective—and biologically effective—way of improving bee stock. Molecular biology is a tool, nothing more, and like most tools it is most effective when it's the right tool for the job and is used as a tool rather than as an end in itself. So don't be confused by all the fancy terminology; if molecular biology is a hammer, and is used to drive in a nail, it will get the job done properly and cheaply. If you use a hammer to tune up your car, however, you'll end up at my mechanic's shop. You'll be considerably poorer for it, and your car still won't run properly.

29. The Business of Research

B Y NOW YOU'VE PROBABLY FIGURED OUT THAT I AM a strong supporter of "bottom line" research with societal relevance. I did not begin my academic career feeling that way, however; in fact, buried deep beneath my exterior advocacy of applied research lies a secret persona I call "Academic Man." My first two undergraduate years were spent at one of the bastions of American ivory tower intellectualized education, the University of Chicago. Our credo was to follow the "life of the mind," and an exciting Saturday night at the U of C was to spend the evening arguing about which university in the world had the best academic library.

I left after two years, transferring to the more proletarian environment of Boston University. Nevertheless, I still retain some admiration for a well-turned academic phrase, and I still value research as much for its own sake as for its problem-solving aspects. While I believe universities should be useful institutions, I also believe the best way to be useful is to train students to think independently and with originality. Universities should not be manufacturing centers for technicians limited in their capabilities to doing only what they are told. Society is well served by university graduates who consider the relevance of their work, but it is not well served by graduates who are limited to performing rote techniques, and that only under someone else's direction.

The issue of how to train students to be original thinkers has become increasingly problematic in modern universities, for a simple reason. The current funding environment requires that our grant applications to government agencies receive matching funds from industry before they will be considered. In other words, researchers need to convince a business, commodity group, or beekeeper/grower organization that our work is important enough to them to provide us with at least half of the dollars we need to conduct that work. The plus side of this funding formula is that it forces us to come up with projects that are of economic value to society. The downside is the danger that university research will become limited to end-user, immediate product development work, and we'll eliminate all the basic research that is the foundation for commercial applications.

This trend is especially troubling for honey bee and pollination research, because beekeeper and grower organizations do not have a lot of money to invest in research. In Canada, the quantity of research potentially beneficial to beekeepers dropped by more than half in 1996, when federal and provincial governments cut budgets dramatically, and it's probably going to get worse, primarily because the beekeeping industry traditionally has not provided funding to support research.

Like my colleagues, I am becoming adept at balancing my mandate to educate students to become original thinkers with the reality that I won't be educating any thinkers at all, original or otherwise, unless industry is willing to foot part of the bill. The new funding reality has directed me to consider different types of projects than I thought about before, and has forced much of my laboratory's research into the fundable mold. I find that I no longer think as much about what projects might be interesting and potentially important to conduct; rather, I confine my thoughts to projects industry might fund.

Although industry funding does establish real limitations on what researchers can do, it is still possible to direct students toward original research that both educates them and fulfills society's orders to do relevant research. Nevertheless, my students' projects have subtly shifted in recent years in the crucible of industry involvement, and they have had to learn the importance of compromising their idealized projects when faced with funding realities. I try to involve my students in grant writing and negotiations with industry so that they can experience firsthand the give-and-take of real-world fund-raising.

I don't know if the quality of my laboratory's research projects has been improved by industry involvement or not, but it has been changed. For example, one of my students wanted to examine the use of the orchard bee *Osmia* as an alternative pollinator for apples. Beekeepers and growers are aware that honey bee colonies are not as available for pollination as they were in pre-*Varroa* days, and both groups are concerned about the shortage of honey bees for commercial pollination. The apple grower organizations we approached for funding were interested, but broke; the beekeepers had mixed feelings, but were not particularly enthusiastic about supporting research that if successful would provide an alternative to the honey bee.

The end result of these negotiations was that my student switched both her approach and the crop. Now, she will be working on blueberries. The last few years have been good for blueberry growers in British Columbia, and they had funds to invest in this project. The scope of the study also changed dramatically. Now she will be comparing the efficacy of four bee species as managed pollinators and testing methods of improving the pollination effectiveness of honey bees by removing pollen from colonies before and during blueberry bloom. In this case, the apple growers' lack of money and the beekeepers' lack of interest forced us to propose a very different study than the one we had originally conceived. Only time will tell whether the blueberry project will prove as worthwhile as the original apple project, but it is clear that industry funding dictated its scope and direction.

Another current project in my laboratory influenced by industry funding involves the pesticide neem. There is considerable interest in neem among beekeepers, because preliminary studies suggest that it might be useful against honey bee pests and diseases. One of my students has begun a project to evaluate the safety of neem for use in bee colonies as well as its effects on parasitic mites and bee diseases. We needed dollars for his project, as for any research, and we negotiated with a number of companies involved in neem sales in North America. One company was very interested but backed out in the end because of legal concerns involving some hotly-contested neem patent issues that are now moving through the courts. We finally concluded an agreement with an Australian company that hopes to supply neem to the North American market, but the "price" of this agreement is that we must use the company's product for tests, and also must negotiate with this group first should any patent or license result from this work. In this case, industry involvement did not directly affect the research we want to conduct, but may have some implications as to when and how widely the results are disseminated if we are successful.

I am about to embark on negotiations for another project, and perhaps the strongest evidence I can provide concerning the impact of industry funding on research is this: I can't tell you much about it. All I can say is that it involves using honey bees as a model system to test certain types of new human drugs, a collaborator from another university is involved, and preliminary research in our laboratories has led us to believe that pharmaceutical companies may be very interested in this project. It also, by the way, is an exciting project because of what we may learn about honey bee biology, and may ultimately provide real breakthroughs in management for honey production and pollination.

You can see the bind university researchers are in today. Society has given us the mandate to educate and to stimulate, to come up with exciting new knowledge using open discourse and free exchange of ideas. At the same time, we are not given funds to fulfill this mandate, but instead are told to turn to industry and accept the limitations on discussion and project direction that come when private business has an interest in university research. Previously, funding came largely from the government, and while industry funding was desirable, it was not required. Today, industry money can be a necessary component of grants, and universities have established university-industry liaison offices, developed model contracts, and engaged patent attorneys to deal with the complex issues that arise when industry funding enters the university world.

I honestly don't know what I think of all this. My laboratory still seems the same to me: my students continue to discuss, argue, and excite each other while pursuing their ideas. They continue to be interested in blending the pursuit of knowledge with service to society by taking that knowledge and developing commercial applications. What I fear the most is that I may be forced to become secretive in order to get funds to maintain my laboratory.

I wonder what kind of discussions are taking place during late-night bull sessions at my old alma mater, the University of Chicago? Do students still wonder about where the best libraries are, or do they trade information about patent rights, license agreements, and rumors of high-level deals their professors are negotiating? We need to remember that "useful" research does not mean "secret," and that a university system in which an increasing proportion of research is tied up in contracts with private industry has become less than a university should be. In the end, society's strongest defense against mediocre research is to encourage students to expand their thinking beyond contract work and into the realm of the truly novel. The ivory tower never was a good model for universities, but the corporate office of a patent lawyer is not a good one either.

30. How Do We Know That?

I HAD AN EXPERIENCE A FEW YEARS AGO THAT DRA-matically changed the way I approach classroom teaching. I often teach a third-year undergraduate course called Insect Biology. One day, I was lecturing about how the insect external skeleton prevents water loss, and how that is particularly important for small organisms such as insects. Suddenly I paused and looked around at the students, who were tediously trying to take down every morsel of knowledge that fell from their erudite professor's lips. It occurred to me that these students were accepting everything I said at face value, without question, and certainly without understanding the scientific process that might lead me to make a statement claiming that insect skeletons prevent water loss.

I put away my lecture notes and told my students to close their notebooks. Then I asked them a very simple question: "How do we know that?" At first, their blank stares made it apparent that they knew very little about the nature of the experimental work that might have led to our knowledge about insect skeletons. Then, however, we began to design experiments that might prove or disprove hypotheses about how insects regulate water loss, and why body size is important. They started to bring in concepts from other courses, to make connections, and even to pose new questions.

I have continued to use this teaching technique, and now often pause in the midst of my lectures to ask "How do we know that?" This question goes well beyond the university classroom, and is one all of us should ask when confronted with scientific evidence about any subject. We are barraged with seemingly endless issues that require us to understand science. The news media are filled with reports about declining salmon stocks, increasing automobile emissions, balancing timber-cutting quotas with wilderness preservation, protecting endangered species, and many more topics for which a clear understanding of the data behind the experts' opinions is crucial to decision making. Yet, nonscientists have both little feeling for how science is conducted and an unrealistic expectation that science can provide simple, unambiguous solutions for complex problems.

I thought it might be interesting to look at a current research project in my laboratory that illustrates how scientists approach a scientific question, and especially how the answers are not always as clear as we would like. This project has been conducted by Dr. Jeff Pettis, a postdoctoral fellow who now is working for the U.S. Department of Agriculture, and Heather Higo, a research technician still in my laboratory, in collaboration with a laboratory headed by Dr. Gene Robinson at the University of Illinois. The project asks a simple question: What happens to worker bees in the queen's absence that induces them to begin rearing new queens?

The first step in any research project is to come up with a hypothesis, an idea to test. Truly novel ideas, the ones that make quantum leaps in knowledge, are extremely rare. More commonly, we take small steps by thinking about things we have recently learned, and putting known ideas together in new ways. In this case, the information we put together came from work we had conducted in my own laboratory and research then being done by Robinson's group.

The queen's presence in a colony prevents workers from rearing new queens, at least until colonies become crowded and pepare to swarm. The main factor inhibiting queen rearing is a pheromone produced in the queen's mandibular glands that is secreted by the queen, picked up by workers, and distributed throughout the colony. Queen rearing begins when the queen is suddenly removed or when colonies become so crowded that pheromone distribution is hindered, and many workers do not come into contact with the queen's pheromone. Thus, it seemed reasonable to suggest that queen pheromone does something to a worker bee that keeps her from rearing queens. Therefore, removing a queen—and her pheromone—should induce some change in the worker's body that stimulates queen-rearing behavior.

Gene Robinson and I began discussing this problem during a visit I made to his laboratory a few years ago, just as he and his students were beginning to study how chemicals in bee brains are involved in foraging behavior. The substances Gene and his students were studying included the neuromodulators dopamine, serotonin, and octopamine (the first two are also found in human brains). We thought that the quantity of one or more of these chemicals might be influenced by queen pheromone, and that removing the queen would result in a change in brain chemical levels leading to queen rearing. If so, we might be able to develop ways to control bee brains, thereby perhaps preventing swarming, improving rearing techniques for commercial queen production, or allowing us to maintain queenless colonies for long periods.

The linking of these concepts between our laboratories illustrates an important point about contemporary science: it is highly collaborative and interactive. Jeff, Heather, Gene, and I continued to discuss this hypothesis at various meetings and through a barrage of e-mail messages that were the electronic equivalent of conversation.

Through this cyberspace chat we were able to design a simple experiment to test our idea.

"Testability" is another important component of scientists' methodology. To test any hypothesis, you need an experiment in which everything is kept the same except for the one factor you wish to examine. For this 1994 experiment, we set up groups of colonies with eight thousand workers each, and made sure each colony had the same amount of brood, stored honey, and pollen. The only thing we varied among the colonies was whether they had a queen or queen pheromone. The colonies were divided into three treatments: (1) queenright (i.e., with a queen) control colonies, (2) queenless colonies to which we added synthetic queen pheromone, and (3) queenless colonies that had nothing added to them. Four and eight days into the experiment we sampled worker bees from each group, dissected out their brains, and shipped them to Illinois where Gene and his crew could determine the levels of the chemicals we were interested in.

The results seemed to be clear. Workers in the queenless colonies without pheromone reared queens and showed significantly reduced levels of all three brain chemicals after four days compared with workers from queenright colonies, which did not rear queens and had higher levels of brain chemicals. This suggested that the three brain chemicals are maintained at normally high levels in the queen's presence, and that removing the queen induces reductions in brain chemicals that cause worker bees to begin rearing new queens. Further, workers in the queenless colonies with pheromone added did not rear queens and had neuromodulator levels similar to those of workers from the queenright colonies. This result suggested that queen pheromone regulates queen rearing by maintaining high levels of brain chemicals.

The next step in "how do we know that" is to question your own results as rigorously as possible, and here we began to get suspicious. First of all, we had expected to find an increased level of one of the three brain chemicals in queenless bees, but instead we found decreased levels of all three. If valid, this result would be highly unusual—and a real breakthrough—but these results were so different from what was known about brain chemicals in other insects that we had to wonder if our results were correct. Second, we saw the drop in neuromodulator levels after only four days, and the levels were unchanged after eight days. Jeff Harris, another researcher in the bee brain chemical field, had noticed an increased level in one of the brain chemicals in bees that were kept queenless for a month, a result that was inconsistent with ours. Finally, our results after four days were simply too clear, too unambiguous, and too novel to trust. We thought we had better make sure we were seeing a real phenomenon before publishing.

A scientific fact must be repeatable, and not easily explained by alternative hypotheses. When we replicated the experiment in 1995, we were disappointed to find that the results we had seen the previous year were just not there. If we fantasized hard enough, we perhaps could see a slight trend here or there that might support the earlier findings, but clearly something was different in 1995. Indeed, if we had done the 1995 work first, and if those had been the first data we saw, we would have concluded that queen pheromone does not influence these brain chemicals, and we would have moved on to look for other things in worker bees that might stimulate queen rearing in the absence of pheromone.

We have continued to discuss, query, probe, and question these divergent findings, and at this point still do not know whether the first or second year's data are more accurate. We conducted further experiments in the summer of 1996 that are now being analyzed, and we hope they will resolve the difference between the data from the two previous years. We are still fairly far from putting something into the textbooks that might explain what happens to a queenless bee.

In fact, the ambiguity of our findings to date is more common in science than the crystal-clear, textbook answers we usually teach in classrooms. This troubles me from a teaching perspective, because students develop an expectation that experiments will lead to answers, whereas they usually lead only to more questions. Collaboration, replication, small steps, ambiguous answers; these are the common currencies of scientific endeavor rather than the sudden insight that leads to a new scientific "fact." We need to teach more "how do we know that" instead of what we think we know, because it is only by asking proper questions that we do eventually come up with at least a small part of the right answers.

31. Consulting

THE BUSINESSMAN WAS FIFTYISH, WITH SILVERY hair and a briefcase that cost more than the annual income of a third world country. Inside the case was wealth beyond imagination, more than a billion dollars in cash and untraceable bonds, so it was not surprising that he looked around carefully before boarding the floatplane. He was heading north, toward an isolated fishing camp in the wilderness of coastal British Columbia. What he didn't notice, however, was that a swarm of bees had landed inside the plane's cabin. As he entered the plane, he felt a sharp prick on his face, then another, and suddenly he was being swarmed by the vicious bees. His last thoughts as he fell into the water, his briefcase sinking into a watery grave, were . . .

OK, I confess: I made that up. Movie plots are on my mind because I recently finished consulting for a television show, and it got me thinking about the various advisory jobs that have come my way as a professor specializing in bees. I am fortunate to work at a university that encourages its faculty members to get involved in off-campus activities. Simon Fraser is the antithesis of an ivory tower, and our faculty tend to be active consultants for business, the media, community groups, government, and so on. I say fortunate for two reasons. First, outside consulting is a way of supplementing our regular income. University professors are not poor by any stretch of the imagination, but our salaries are not bloated either, and consulting work can be a very welcome addition to our paychecks. In addition to the income, consulting work provides a fascinating glimpse at how people in other professions make their living, and at the various perspectives with which they view bees.

My first consulting job was not at all lucrative, but it did have a great impact on my future. I had just completed my graduate research on Africanized bees, and on my way home was invited to the island country of Trinidad-Tobago to advise the government on how to cope with these bees. The Ministry of Agriculture was concerned that swarms of Africanized bees might colonize the country by flying across the nine-mile strait separating Trinidad-Tobago from the South American mainland, which is exactly what happened a few years later. The beekeepers and government officials I met were wonderful hosts, taking me all around the islands and showing me beekeeping operations, tourist sites, marvelous nature preserves, and incredible beaches.

This was heady stuff for a young graduate student, and for the first time I realized that the work I did might have value to someone outside the academic world. I came away from Trinidad-Tobago feeling that my future responsibilities might go beyond publishing papers in obscure academic journals, and that research could be important for beekeeping.

The consulting jobs that have come my way since have diversified over the years as I have become more established in my field. My consulting repertoire now includes legal work, and I am occasionally called on to testify in cases involving bees. I can say unequivocally that lawyers have a very different perspective on bees than beekeepers or governments have. To the legal profession, bees are lean, mean income machines. Considerable money can often ride on a bee case, and an expert witness may make the difference between a tiny award and one in the hundreds of thousands of dollars.

For some reason, the cases I have worked on have all been in the area of workers' compensation. One case was especially interesting because I was able to shift the blame from bees to wasps, and help a poor widow in the process. Her husband, a long-haul truck driver, was stung while carrying a load of wood shavings. He radioed in that he had gotten stung, and that was the last anyone knew until a passing driver found him dead at the wheel twenty-five minutes later. His widow applied for about $500,000 in compensation, but the government-run board turned her down, claiming that the man was no more likely to be stung than an average member of the traveling public, and was therefore ineligible for compensation according to the applicable laws of British Columbia.

I was brought in as an "expert" on bees, and quickly earned my fee by pointing out that there was no evidence a bee was even involved. I argued that the man was most likely stung by a wasp attracted to the wood shavings he was hauling, with the wasp wanting the shavings for pulp to build its nest. This was a key point because it meant that the man was at higher risk of attracting a stinging insect and being stung than an average person driving down the road would be. The board bought the argument, the widow received her compensation, bees were exonerated, and both the lawyer and I were enriched by the process (the lawyer much more than I, unfortunately).

I also was able to help the image of bees on another consulting job, this one for an edition of the Magic School Bus books. This series is a spin-off from the public television show that has a teacher, Mrs. Frizzle, shrink her class down to miniature size in a school bus and take them on weekly science adventures. The bee book involved shrinking the class and driving the bus into a real bee hive, teaching kids about bees, honey, and pollination in the process.

This job was fascinating because I got to think about what it would be like for a kid to really enter the world of bees, and also to observe the writer and artist work at creating a vision of bees that would both inform the kids and grab their attention. I also did beekeepers a great service. The term *botulism* appeared in an early draft, but the authors fortunately followed my forceful advice to remove that dreaded word from the book.

Another call came in a few years ago from a novelist who wanted to use bees as part of a story he was writing. His idea was to have the president of the United States be allergic to bee stings, and have him visit a farm with bees on it. The morning of his visit, a terrorist sneaks into the president's bathroom, substitutes deodorant containing alarm pheromone for the president's regular deodorant, and you can imagine the rest.

I think I dissuaded the novelist from this farfetched plot by proposing an equally farfetched one. In my concept, the president was visiting a farm with bees on it, and was offered a banana for lunch. Bananas naturally contain iso-pentyl acetate, a component of the honey bee alarm pheromone, so when the president bites into the banana, he attracts bees, gets stung, and dies, with no obvious trace left by the terrorist. For some reason the novelist didn't like my suggestion—or even his own, apparently. At least, I've never encountered a book in print that includes bananas or deodorant as part of a terrorist attack on the president.

The television and movie industry is the most fascinating business to consult for, and also the most controversial, because these media look for the most lurid ways of depicting bees. The most recent made-for-TV bee movie (for which I was most definitely *not* the bee-wrangling consultant) had a gargantuan swarm of killer bees take over a California town and terrorize a family in their home for most of the film. I avoid working on those deep-stinging, killer-bee-as-terrorist movies, but sometimes an interesting job comes along that depicts bees from a different perspective.

I recently did some work for the *X-Files*, a television show whose basic premise is that aliens are here on earth, the government knows about it, but the whole thing is hush-hush. The star of the series has dedicated his life to exposing the conspiracy, and in one episode bees enter the picture. In this story, the bees are the good guys. I am not allowed to say more about the plot, but I can tell you that it involves an amazing "interior of a hive" set, that no expense was spared to put the bees into the action, and that you clearly get the idea that bees sting only the bad guys. I had never considered how aliens coming to earth would use bees, or indeed what bees from another planet might be like, and that certainly was a mind-expanding experience!

Consulting work is interesting, but it also presents numerous dilemmas, and I have had to develop financial and ethical criteria to use in considering each job offer that comes along. Some questions that have come to mind include: Should I do "killer bee" movies? Should I charge beekeepers a fee for my advice? If a lawyer asks me to be an expert witness in a case against a beekeeper, should I accept? These are knotty questions, and anyone offered a consulting job involving bees should give them considerable thought before making a decision.

I've resolved these issues with three simple rules: I don't do killer bee movies, I don't charge beekeepers or growers for advice, and I won't advise on a legal matter in which my expert opinion might harm a beekeeper. Fortunately, my simple rules have yet to be put to the test. I haven't been offered work on a killer bee movie or on a legal case against a beekeeper, and I doubt that any beekeepers or growers would pay for my advice anyway. Now, if I could just sell that plot to a movie company . . .